The Technological-Industrial Complex and Education

Sydney Marie Simone Curtis
Victoria Desimoni • Max Crumley-
Effinger • Florin D. Salajan
tavis d. jules

The Technological-Industrial Complex and Education

Navigating Algorithms, Datafication, and Artificial
Intelligence in Comparative and International
Education

Sydney Marie Simone Curtis
Dallas, TX, USA

Max Crumley-Effinger
Emerson College
Boston, MA, USA

tavis d. jules
Loyola University Chicago
Chicago, IL, USA

Victoria Desimoni
Arizona State University
Tempe, AZ, USA

Florin D. Salajan
North Dakota State University
Fargo, ND, USA

ISBN 978-3-031-60468-3   ISBN 978-3-031-60469-0   (eBook)
https://doi.org/10.1007/978-3-031-60469-0

This Palgrave Macmillan imprint is published by the registered company Springer Nature Switzerland AG.
The registered company address is: Gewerbestrasse 11, 6330 Cham, Switzerland

Paper in this product is recyclable.

CONTENTS

LIST OF TABLES

Artificial Intelligence in Comparative and International Education in the Age of the Anthropocene

Abstract This chapter provides a brief introduction to comparative and international education (CIE) as well as a concise review of artificial intelligence (AI). We explain how AI has began to permeate every facet of education and consider its implications for CIE. We examine what a human-centered approach to AI should look like. We then discuss the researchers' positionality that has guided the construction of this book. The core research question that drives this book is identified. We present an outline of the six chapters that make up this book.

Keywords Artificial intelligence • Human-centered approach • Processing-tracing • Comparative and international education

INTRODUCTION

Quick change breeds panic. The discourse around artificial intelligence (AI) often focuses on its potential dangers: algorithmic bias and discrimination, the mass destruction of jobs, information fragmentation leading to islands of disjointed sets of truths, and even, some say, the extinction of humanity. While some scholars fret about these dystopian scenarios, others focus on the potential rewards. Nevertheless, commentators all agree that AI is working at deep and complex levels in ways that radically redefine

© The Author(s), under exclusive license to Springer Nature   1
Switzerland AG 2024
S. M. S. Curtis et al., *The Technological-Industrial Complex and Education*, https://doi.org/10.1007/978-3-031-60469-0_1

what it means to be human. In comparative and international education (CIE), which is the scientific study of education from a comparative perspective and the study of foreign educational systems, AI can release creative disruption and transform the field. It may possibly help CIE researchers solve some of humanity's most complex and thorniest problems. Suppose AI is used in an ethical and just way in CIE. In that case, it has the potential to radically reshape the field by accelerating and turbocharging the pace of scientific research and discovery,[1] especially in areas such as international development, education in emergencies, refugee education, peace and conflict study, educational policy, etc. The question guiding this book is *how CIE scholars and practitioners can use AI to decolonize research and practice in CIE.* In this way, we use a decolonial approach to expand the scope of discursive analyses and practices to incorporate a critique of the coloniality of power by examining how modern/colonial constructs of epistemic authority are succored.

So, what, then, is AI? The concept of AI, which was first proposed at the Dartmouth Conference in 1956, has far-reaching consequences in light of the sweeping changes it has caused in recent years (referred to as "the big leap"). The search for a standard definition of AI is still ongoing, and only by defining the concepts "artificial" and "intelligence," as well as the relationship between them, can the term "artificial intelligence" be truly understood. *Artificial* is viewed as something that is not naturally produced, that is, something produced by human beings. In the context of intelligence or systems, artificial has come to denote something that is fabricated or simulated through human-designed processes. *Intelligence* is multifaceted and often viewed as the ability to obtain and harness knowledge, problem-solve, adjust to new circumstances, and learn from experience. Intelligence involves reasoning, understanding, learning, planning, and problem-solving. As such, *artificial intelligence* is the development of computer systems to perform tasks typically done by humans and require human intelligence. In this way, AI uses computer systems to mimic and stimulate human intelligence with the aid of data, algorithms, and computational power. For us, AI is defined as using a machine-based technique and learning platform with algorithmic power for making predictions, diagnoses, recommendations, and decisions and performing and

---

[1] At the time of the writing of this book, several types of AI have already been deployed in other scientific disciplines (such as wildlife conservation, neuroscience, pure mathematics, fundamental physics, linguistics, and research methods) (see Economist, 2023).

mimicking tasks requiring human-like intelligence, cognition, and the ability to adapt to the immediate environment. AI is an umbrella term that subsumes methods, algorithms, and systems that learn from data (data science, statistical learning, and machine learning[2]) and aim to create machine intelligence or machine learning that can perform tasks such as observation, analysis, and assumption (for example expert systems, probabilistic graphical models, and Bayesian networks). While AI is associated with computers, there is a movement to see it as much broader than this with the onset of embedded computers, sensors, and other emerging technologies. In other words, AI refers to the capacity of a computer program to learn and reason or "the theory and development of computer systems able to perform tasks normally requiring human intelligence" (Jobin et al., 2019, p. 389). Any task performed by a program that would typically require human intelligence is classified as AI. Coppin (2004) suggests that AI is premised upon machine adaptability to new situations, dealing with emerging situations, solving problems, answering questions, devising plans, and performing various other functions that require some level of intelligence typically evident in human beings.

Since AI has been adopted and permeated within various areas of "the education sector or departments in educational institutions, it has impacted and improved efficiency, global learning, customized/personalized learning, smarter content, and enhanced effectiveness and efficiency in education administration, among others" (Chen et al., 2020, p. 75265). However, the rapid development and expansion of AI also necessitate a critical moment to step back and reflect on the implications of AI in CIE and for social justice in education more broadly. This book explores how artificial intelligence can be more accessible, humane, and just, particularly through its applications in CIE. This book will focus on how AI functions in CIE from both conceptual and practical lenses. Essential questions related to this theme that will later be explored include: How should we use AI to better humankind? What are some suggested approaches to

---

[2] Machine learning uses computation data to perform complex training and learning, such as face unlock, speech recognition, natural language translation, and virtual reality. Deep learning is a subfield of machine learning in which "layered representations are (nearly always) learned via models called neural networks structured in literal layers stacked on top of each other" (UNESCO, 2019a, p. 9).

decolonizing AI in CIE? And what are the barriers and opportunities to promote social justice in using AI in CIE in the digital society era?[3]

The big leap is taking place within the Fourth Industrial Revolution, an era in which the scientific-technological-industrial complex has altered educational systems, and all human activities cannot be separated from the use of Information Technology (IT). In this digital era, IT has emerged as both a tool and mechanism that aids in human life activities. Rapid advances in Big Data[4] (which combines data-management technologies) and AI, the Internet of Things (IoT), and other information and communication technologies (ICT) have profoundly impacted all aspects of human society. While ICT involves the storing, processing, or transmitting of information units, AI (which includes many automation capabilities) is developing rapidly, so it is almost impossible to keep up with new technological advances. However, even as we race to keep up with the latest developments in AI, we can still endeavor to trace and contextualize its rise to prominence in education.

The new information landscape of rapid digital transformation and digitalization will transform CIE research. We have to consider how AI is applied to CIE and the ways in which we can mitigate misguided errors. Numerous countries are currently developing AI policies and governance mechanisms with a focus on making these guides and principles accessible to the average person. AI is becoming more accessible with the advent and continuous development of text-to-text, text-to-image, and text-to-music generative AI models. These new models can provide CIE researchers with the next generation of tools to assist them in solving some of the challenges of the Anthropocene. However, we have to be careful because, with the advent of the democratization of AI, it has the potential to misuse and abuse information. As such, we need to ensure that the field develops realistic ethical guidelines for information usage as well as have an inclusive approach. Some have called for a human-centered approach to AI that involves "properly gather[ing] training data (ensuring it is free of bias) and to scrutinize the models' algorithms for potential bias" (Kishimoto et al., 2024, p. 3). In its Ethics Guidelines for Trustworthy AI, the Independent

---

[3] The digital society/economy is defined by the movement away from the post-Internet society. It is driven by an exceptional socioeconomic environment in which electronic communications prevail, and cybersecurity has increased importance.

[4] Big Data allows educational institutions to sort, manage, and process vast amounts of data to gain insights. Big Data is defined by the characteristics of the 3Vs: volume (amount of data), variety (diversity and types), and velocity (speed to generate new data).

High-Level Expert Group on AI established by the European Commission states that a human-centric approach to AI should be rooted in "respect for human dignity, in which the human being enjoys a unique and inalienable moral status" (European Commission, 2019, p. 37). Such an approach is based on the idea that AI should be centered around humans and be designed by humans, for humans and with humans, with the aim of promoting human flourishing (see Vallor, 2024). In other words, "human-centered AI really is about the alignment between the humans driving the technology and the humans standing at the center of where the technology is being driven" (Vallor, 2024, p. 14). Vallor (2024) notes that current AI models are not human-centered, and we are now forced to "accommodate" these technologies that have been forced upon us. In fact, educators and researchers were not asked how AI could help them teach and do research better; instead, generative AI platforms such as ChatGPT were foisted upon them. A human-centered approach to AI begins by asking: What do we need from AI? How can AI shape our values? And what role can Indigenous knowledges play in shaping AI? Instead, what we have gotten, according to Vallor (2024), are scaled up technologies that have been scaled up as big as we can get them "and pour as much computing power into them as we could [...] And what we get is something that we then have to cope with as opposed to something designed by us, for us, and to benefit us" (p. 16). Because non-human-centered models are being spat out in the name of progress, research must adapt to AI tools as opposed to AI tools adapting to research. Therefore, tools such as value-sensitive design and participatory research design are essential facets of human-centered AI in that these designs are built around asking critical questions. If we are not designing educational AI with the researchers in mind who will use the technology, then we have already failed. For AI to be integrated appropriately into CIE research and be human-centered, researchers must have a role in the process of conceiving, designing, testing, deploying, and iterating these technologies.

OUR POSITIONALITY

We all received most of our university education in English and predominantly employ English as our academic language in the United States. Our positionality points to Moosavi's (2020) reminder for "Northern scholars who need to recognise how we are privileged by coloniality and even implicated in its enduring structures of inequality" (p. 333). For us, engaging in decolonial work is part of our broader reflexive reckoning with our positionalities and professional obligation found in our scholarship and praxis. Across this book, we employ a reflective approach grounded in decoloniality and social justice as a way to shift our power and privilege. In doing this, we position ourselves and our work within the transformative (critical) paradigm, which is a Western approach rooted in social emancipation values, engaging in a decolonial praxis and standing in solidarity with oppressed and vulnerable peoples. Therefore, our work utilizes a critical systemic approach that considers theory and practice that are informed by our ideological and ethical stances and, in turn, advises our methodological pluralists' bearing (Cram & Mertens, 2015). Given the hierarchical relation of power that privileges the production of academic over local Indigenous knowledges, we agree with Mora and Diaz (2004) that "the entire research endeavor must be participatory in nature in order to produce qualitatively different research that is based on community-identified problems and needs" (p. 24). Within our respective contexts, we strive to ensure that our research practices engage with the injunction that "decolonization is not a metaphor" (Tuck & Yang, 2012) or what Fishel (2017) calls "metaphorogenesis" which draws attention to how the world is shaped through metaphorical constructs.

OUTLINE OF THE BOOK

In Chap. 2, we delve into the rise of the technological-industrial complex and the transition from Education 1.0 to Education 4.0, highlighting the ethical and just use of AI in CIE. Whereas Education 1.0 was characterized by passive learning and teacher-centered instruction, we show that Education 4.0 emphasizes personalized, collaborative, and technology-integrated learning experiences. The integration of AI and advanced technologies in Education 4.0 aims to prepare students for the demands of the education intelligent economy. The COVID-19 pandemic has accelerated the adoption of online teaching and learning, leading to a rapid advancement of AI in education. This transition reflects the shift towards a digital society and the increasing influence of technology in all aspects of human life. We also highlight the historical colonial roots of CIE and the potential biases and limitations of AI; we address the impact of colonization and modern technological innovation on AI and digital technologies. We argue that these technologies are exacerbating existing inequalities and perpetuating colonial power dynamics in the form of colo-niality. We advocate for the decolonization of AI to address these issues and ensure that AI benefits all individuals and communities. We note that a decolonial lens is essential for understanding the power dynamics and inequalities inherent in the social and technological fabric and call for diversified data and data teams to prevent the reinforcement of historical inequalities.

In Chap. 3, we explore the historical and cultural roots of AI, tracing its origins from ancient mythology to modern science fiction, and discuss its portrayal in popular culture, including in iconic science fiction series and films such as Star Trek, Star Wars, and Terminator. We draw attention to the real-life development of AI, from early programs like ELIZA and the General Problem Solver to more recent advancements such as IBM's Deep Blue, while also addressing the skepticism and challenges surrounding AI, including the limitations of expert systems and the need for AI to replicate human learning processes. Emphasis is placed on the critical examination of the role of AI in education while drawing attention to the philosophical and ethical implications of AI and its relationship to human intelligence. The chapter also addresses the different types of AI, including Narrow AI, General AI, Super AI, Reactive Machine AI, Limited Memory AI, Theory of Mind AI, and Self-Aware AI, and their potential implications for educa-tion, such as enhancing student learning, providing personalized learning experiences, and automating administrative and assessment tasks. We raise ethical concerns about the use of AI in education, including issues of com-mercialization, cultural bias, and the impact on human rights and demo-cratic values.

Chapter 4 provides an overview of the ethical considerations surround-ing the use of AI in CIE by discussing the potential risks and benefits of AI in the field and addressing concerns about digital coloniality, equity, and access. Given that AI in educational research and practices offers sev-eral advantages, including access to diverse data, promoting inclusivity and global collaboration, enhancing creativity, minimizing bias in data analy-sis, and breaking down language barriers, we call for the development of ethical guidelines to govern the responsible use of AI in CIE, highlighting the importance of ensuring that AI technologies are designed and imple-mented in accordance with human needs and values. Since the field of education has limited control over the development of AI, and AI is rooted in colonial logics, we also underscore that the unequal access to AI and its potential to perpetuate educational inequalities must be considered. Therefore, the challenges of incorporating AI in CIE include the potential marginalization of alternative knowledge and the erasure of local considerations.

Chapter 5 discusses the methodological implications of using AI in CIE research, exploring the impact of AI on research methods and methodologies in CIE, as well as the potential challenges and ethical considerations associated with its use. It emphasizes the need for ethical and community-centered research practices, such as participatory research and action research, as alternative methods for community-led research. In stressing the importance of critical literacy, community involvement, and obtaining informed consent when using AI tools, we suggest that there is a need for more empirical studies with direct community involvement to examine the impact of AI tools on research practices in the field. The use of AI in CIE research presents ethical challenges, particularly in data collection, analysis, sharing, and reporting. However, AI tools, such as conversation robots and natural language processors, can assist researchers in synthesizing data, recalling historical facts, and expediting literature reviews. However, CIE researchers must be transparent about the use of AI in their studies and consider the ethical implications of AI-assisted research.

Chapter 6 provides an overview of AI governance in education, with a focus on regulatory responses and emerging global scripts. It discusses the shift of AI development from academia to private industry and the increasing interest from policymakers in addressing the opportunities and threats posed by AI in education. The chapter examines various AI governance frameworks from the European Commission, Türkiye, China, UNESCO, and the US Department of Education, highlighting their approaches to AI in education and emphasizing the importance of understanding the diverse purposes of education and the local, regional, and national contexts in shaping AI governance. We underscore the importance of considering equity, decolonization, and epistemic inequity in the development and implementation of AI systems in education. In emphasizing the need for ethical, equity-focused, and decolonial approaches to AI-influenced education, we suggest that CIE scholars can play a role in shaping governance frameworks in this area.

We conclude the book by returning to the concepts of decoloniality and pluversality.

CHAPTER 2

The Rise of the Technological-Industrial Complex and Education 4.0

Abstract This chapter examines the rise of the technological-industrial complex, which is a multifaceted system encompassing a wide range of technological domains and how AI has revolutionized our thinking. It explores the nexus between creative disruption and disruptive innovation and their possibilities for the field of CIE. The linkage between Education 1.0 to 4.0 and Web 1.0 to Web 4.0 is studied in the context of industrial and educational developments. It shows that educational developments are tied to the development and expansion of the Web. It then discusses that CIE was founded as a colonial project and continues to function as such, and in the age of advancement in AI, CIE needs to be decolonized as it enters the Fourth Industrial Revolution. The ways in which AI needs to be decolonized are examined so that it may be provided in a just and equitable way that benefits all of humanity.

Keywords Technological-industrial complex • Decolonization • Creative disruption • Disruptive innovation

INTRODUCTION

The rise of the "technological-industrial complex" (TIC), defined as the interdependence of networks, has implications for CIE. In essence, the TIC is an unintended byproduct of the so-called Fourth Industrial

© The Author(s), under exclusive license to Springer Nature
Switzerland AG 2024
S. M. S. Curtis et al., *The Technological-Industrial Complex and Education*, https://doi.org/10.1007/978-3-031-60469-0_2

Revolution, and we argue that the "educational intelligent economy" (Salajan & jules, 2020) and technology complexity drive TIC as the boundaries of nature and culture recess. In this context, educational intelligence encompasses any individual or systemic data that can be recorded and disseminated while improving educational systems. The educational intelligent economy is "premised on the exponential production of digital data to measure, analyze and predict educational performance in comparative perspective" (Salajan & jules, 2020, p. 1), and technology complexity is the level of interdependence between the physical components of technology. As a data-driven culture continues to emerge, there is a growing demand for educational intelligence emphasizing interoperability to link disparate data assets and enable informed decision-making. The exponential growth and consolidation of corporations, the technological establishment and government bureaucracy after the post-Cold War period and the rapid development of communication technologies and infrastructure created the TIC. This did not occur by happenstance; it was a constitutive process that sought to integrate the elite power structure and the political system, thus allowing the technology industry and its financial resources to disproportionately influence the priorities and direction of modern-day educational institutions. TIC, which can be seen as a continuation of the military-industrial complex (MIC) that emerged in the United States during the Cold War, is a complex and multifaceted system encompassing a wide range of technological domains defined by the close collaboration between industry, governmental agencies, and organizations to develop, produce, and disseminate technology-related products and services. TIC is driven by technological innovation through public-private partnerships, a critical factor for educational advancements such as improving educational outcomes and increasing access to education. Through interactive whiteboards and tablets, TIC has manifested in online learning platforms, learning analytics, and adaptive educational software to personalized learning systems. Therefore, TIC has consequences for how we work and learn, and it is hard to separate the hype from fearmongering and informed concerns.

While AI vows to revolutionize education, it is not the first technology to promise a panacea. In fact, we have lived through several episodes of disruptive innovation (Christensen, 1997; Christensen & Raynor, 2003), where new products and services have altered the existing market, replaced established competitors, and fundamentally changed how things are done. If AI is viewed as a disruptive innovator for education, this implies that global educational systems have failed to stay attuned to market shifts. Disruptive innovation in education involves the introduction of new and innovative models and technologies that challenge traditional educational practices. It aims to enhance existing learning opportunities, create new ones, and improve overall student outcomes. It is imperative to realize that education is inoculated from disruptive pressures. As our world is evolving at an unprecedented pace, so is the need for education systems to adapt to the changing reality. Hence, it is crucial to acknowledge that these challenges need to be recognized and addressed with creative solutions, innovative teaching methods, and flexible strategies to ensure that education stays relevant and practical. Thus, disruptive innovation holds that when certain types emerge based on market and technological changes driven by technological complexity, entities unable to keep up will fail. Hence, new technological innovation is either viewed as a threat or an opportunity. Christensen et al. (2016) remind us that "incumbents value sustaining over disruptive innovations because they prioritize their existing customers; they may not be concerned with nascent disruptive threats that exist within largely separate resource networks" (p. 9). While the field of education is not susceptible to new entrants displacing incumbents often, this implies that education cannot resist the forces of disruption, and it will have to find a way to accommodate new types of technological innovation. Perhaps what is striking about AI as a disruptive force for education is that it must grapple with the sheer speed at which it is occurring. This is because the twenty-first-century student has access to vast amounts of information, educational technology, and both traditional and online learning platforms. Educational innovations play a crucial role in promoting growth and sustainability in the field of education. Disruptive innovation in education driven by AI can promote better learning outcomes, student engagement, critical thinking, and problem-solving skills. Not so long ago, in education, we all thought that Massive Open Online Courses (MOOCs) would change the business of education. MOOCs were touted as the next wave of technological innovation to revolutionize higher education by offering online courses that were affordable and flexible ways to learn educational skills and did not displace the traditional classroom structure in the same way that Airbnb displaced the hotel industry. It was seamlessly adopted into the existing higher education infrastructure, with several university and business partnerships being created to facilitate its usage. Nevertheless, if done correctly, disruptive innovation can harness the power of new technological innovation.

Similarly, creative disruption, which is the intentional use of innovation to disturb the status quo, promised to revolutionize educational systems in the past. This involved breaking away from conventional thinking and challenging the prevailing norms to engender change. This was the shift from direct instruction to using Learning Management Systems (LSM) in education. Creative disruption challenges the traditional understanding of education and recognizes that students learn in different places (online learning), the value of online cooperation (chat-based collaboration), that students learn at their own pace (AI-guided learning), and ensures safe ways to expose students to real-world situations (virtual and augmented reality). The potential of creative disruption to transform CIE is enormous in that it can foster different learning environments that encourage curiosity and critical thinking through interactive online platforms, virtual reality simulations, or even AI-driven personalized learning. Given the multifaceted dimension of globalization that has impacted national educational systems, creative disruption involves reimagining teaching methods and embracing technology to enhance learning experiences.

In CIE, the disruptive innovation and creative disruption brought about by AI present new possibilities and challenges to studying the world's educational systems in comparative perspective in that they can decolonize educational technologies and impact CIE's academic and practitioner life in all of its facets (teaching, research, international collaborations, international understanding, and development work). AI has brought progress, and it can also bring disruption. As we navigate how AI rapidly transforms our world, we might also ask how best we can use AI to help students, researchers, and practitioners achieve educational outcomes. This question is even more pertinent in CIE, given that this field benefits from studying education through a comparative lens. To begin answering this question, we examine the challenges and opportunities that AI poses to the field of CIE and use an equity and decolonization lens to analyze the implications of AI. The following sections of this chapter will overview the rise of AI in the context of other movements in technological advancement that have permeated education in recent history, specifically the growth of the educational intelligent economy and the transition from Education 1.0 to Education 4.0. Next, we briefly trace coloniality in CIE and its implications for AI. We then discuss how, as machines develop intuitive ways to amplify human intelligence, we should think about how we can decolonize AI in the era of the Anthropocene.

INDUSTRIAL AND EDUCATIONAL TECHNOLOGICAL DEVELOPMENTS: FROM EDUCATION 1.0 TO 4.0

Schwab (2016) posits that we have undergone three industrial revolutions and are currently in the fourth, all of which can be classified as Eurocentric/ Western revolutions. The First Industrial Revolution occurred circa the seventeenth hundreds and used water and steam power to mechanize production. During this time, many Western countries developed compulsory schools. The Second Industrial Revolution, originating in the mid-nineteenth century, was driven by the creation of mass production and the rise of consumerism. Schooling grew during the Second Industrial Revolution. Many of the Western standards we have today were developed during this period, and the classic model of teaching based on rote learning was developed. This period saw the rise of the "knowledge economy" built out of the "Information Age" (first coined by Drucker in 1969) and focused on developing human capital. The Third Industrial Revolution, or the digital revolution of the mid- to late twentieth century, revolved around the movement from analog electronic and mechanical devices to pervasive digital technology, and it was driven by automated production and premised upon the use of electronics and information technology (semiconductors, mainframe computing, personal computing, the Internet, and the Web). Schooling, particularly access to university education, increased as access to information became freer and immediate. Finally, the beginning of the twenty-first century saw the rise of the Fourth Industrial Revolution which brought with it Web 4.0, a technological revolution and a digital transformation that can fundamentally alter the way we work, live, and relate to each other. The educational intelligent economy drives this new era. Nevertheless, it can potentially transform education as AI is central to it. As such, the transition from Education 1.0 to 4.0 correlates to Web services 1.0, 2.0, 3.0, and 4.0 development.

Under Web 1.0, websites were based on HyperText Markup Language (HTML) code, and interfaces were unintuitive. At the same time, the content was managed by specialized creators, leading to internet users becoming passive and unable to comment on content (Huk, 2021). In many ways, Education 1.0 looked and functioned like Web 1.0. Education 1.0, which is often viewed as "an *essentialism* or *instructivism* teaching and learning philosophical orientation," is based on the three Rs—"*receiving* by listening to the teacher; *responding* by taking notes, studying texts, and doing worksheets; and *regurgitating* by taking the same assessments as all other students in the cohort" (Gerstein, 2014, pp. 83–84 emphasis in original). Essentialism focuses on instilling students with the most basic or essential academic knowledge based on the teacher deciding what students should learn. At the same time, instructivist learning, a teacher-centered model, views knowledge as existing independently from the learner, and has to be transferred from the teacher to the student. In Education 1.0, the student is a passive recipient who cannot comment, share, or interact with the content, the system is teacher-centered, and technology is typically excluded from the classroom. Under the traditional classroom approach inherited in Education 1.0, teaching is based on a one-size-fits-all lecture-based approach. Learning activities are based on textbooks, and the industry views graduates as assembly line workers. Under this didactic format, the teacher is the gatekeeper of information and provides students with content and knowledge that they deem necessary. Thus, students were the consumers of information, and although they would engage in activities, these were mainly done in isolation. Accordingly, technology was based on the passive procurement of online knowledge. In this era, Virtual Learning Environments/Learning Management Systems (LMS/VLE) emerged to facilitate creating, managing, and delivering course material, and educators started attempting to enhance face-to-face learning with the Web (Salmon, 2019). These systems would facilitate the transmission of instructions and electronic resources to enhance student learning within a collaborative setting while empowering instructors to concentrate on crafting meaningful educational experiences (Turnbull et al., 2021).

By 2005, the democratization of the Internet led to the emergence of Web 2.0, also known as the "read-write" Web. This evolving platform allowed users to control and edit their content, blurring the boundaries between creators and recipients. The norm became striving towards collective intelligence, exemplified by platforms such as Wikipedia. During this period, several novel platforms emerged, including blogs, wikis, sharing sites, music, images, videos, and more, allowing users to collaborate and interact with each other. As Salmon (2019) notes, the significant change from Web 1.0 to 2.0 was not the technology itself, but the way in which it was used. In the new millennium, integrating technology into the educational process marked the advent of Education 2.0. Educators and learners began leveraging technology fundamentally, moving away from just accessing information and content to directly interacting with it by commenting, remixing, and sharing it via social networks. Under Education 2.0, the focus became the three Cs: communicating, contributing, and collaborating. Teachers and students interacted more, and education took place "between a student and a teacher, a student and a student, a student and content, as well as a student and an expert" (Huk, 2021, p. 38). The emphasis was on the "humanistic roots" (Gerstein, 2014), where the human element was placed at the center of learning, and technologies were used as a way to enhance traditional approaches to teaching. While "blended learning" (Khe Foon & Kwan, 2018) became popular, teachers still facilitated learning under a constructive teaching philosophy, where the principles of active, experiential, authentic, relevant, and socially networked learning experiences were built into the class or course structure. In Education 2.0, knowledge construction was emphasized instead of knowledge reproduction, and learners were provided with real-world scenarios that encouraged collaboration and social negotiation. During this period, as mobile technologies and better integration became possible, there was a movement away from "information transfer" towards "information assimilation" (Salmon, 2019), premised upon application and learning together. Such a transition paved the way for the development of technological platforms based on academic integrity and plagiarism detection, and the subsequent rise of MOOCs.

A focus on the openness of knowledge coupled with the rise of AI is what distinguishes Web 2.0 from Web 3.0 or the "Semantic Web"[1] in that Web 3.0 can process the content of websites using "new programming languages which categorize and manipulate data in order to enable the machines to understand these data and the phrases which describe them— obtaining information from an increasing range of sources, from inaccessible applications—creating and sharing all types of data across all types of the Web via all kinds of devices and machines" (Huk, 2021, p. 39). Salmon (2019) maintains that "the evolution, maturing and integration of the Web to describe applications which are capable of 'talking to' and exchanging data automatically between each other" (pp. 99–100) is what defines Web 3.0. With the advent of AI, computers have become more intelligent and can now understand information and deliver faster and more accurate results. Thus, information is delivered in a "targeted and user-preferred manner" to a user who "seeking specific information in the Web will receive (as a result) well-selected references" (Huk, 2021, p. 39). Web 3.0 saw the development of integrated web environments driven by "bots," i.e., computer programs that behave like human beings who can recognize

---

[1] The science of machine comprehension of text is called semantics.

and sequence data in a human-like manner, which made Web 3.0 omnipotent and super-integrated. Alongside these technological advancements, Education 3.0 emerged, characterized by the ubiquitous utilization of the user-generated Internet. Schools and universities were no longer the sources of truth and learning, and their role had to shift from transmitting to facilitating learning as knowledge. These advancements facilitated learners' access to self-sourced information, electronic learning opportunities, and communication platforms with peers and instructors. The educational landscape became more networked, with students possessing direct access to diverse sources of knowledge, rather than relying solely on the traditional teacher-student interaction. In Education 3.0, learners are central in creating knowledge artifacts, and emphasis is given to a different set of three Cs—connectors, creators, and constructivists—where the focus is on learning rather than doing (Gerstein, 2014; Huk, 2021). Thus, students of Education 3.0 "live online" and "learners are highly autonomous and self-determined, and emphasis is placed on the development of learner capacity and capability" (Gerstein, 2014, p. 92) as students move to a growth mindset. With the gradual move towards Education 3.0, students become more conscious of how they learn and are taught.

Education 4.0 is linked to Web 4.0 or the "symbiotic web," which is driven by Globalization 4.0 and Industry 4.0—which are sustainable production processes and integration of these processes with digital media. It is considered symbiosis because it links artificial and human intelligence together and dictates how they interact and gain experiences from each other (Salmon, 2019). In this way, AI represents the ubiquitous technological application in all areas that we previously thought only humans could undertake. Thus, Web 4.0, with its integration and technological transformation driven by AI, has led advances in IoT, machine learning, drones, autonomous vehicles, nanotechnology, biotechnology, precision medicine and genomics, advanced materials, smart grids, robotics, and Big Data all of which have finally trickled down into education. The difference between Education 4.0 and Education 3.0 is "the massive ubiquitous connectivity and the *symbiosis* between humans and machines and, consequently, our inability to understand or predict what this might mean in the long term" (Salmon, 2019, p. 102, emphasis in original). As such, Education 4.0 is a response to the rapidly growing intelligent economy, a central component of the Fourth Industrial Revolution. Education 4.0 focuses on transforming the future of education through advanced technology and automation by altering the learning process by incorporating technology into the curriculum to enhance the students' learning experience. As Fisk (2017) notes, under Education 4.0, learning takes place anytime, anywhere; learning is personalized; students have a choice in their learning; project-based learning is centered; students are exposed to hands-on learning; students are expected to apply theoretical knowledge to data interpretation; students are assessed differently; students are considered in designing and updating the curriculum; and students have more independence in their learning. Thus, Education 4.0 merges the digital world with the analog world and "focuses mainly on the new qualification requirements and on adapting curricula to Industry 4.0" (Huk, 2021, p. 37).

In Education 4.0, teaching is discussion-based and driven by the personalization of education, where teachers are viewed as mentors and facilitators. This system continues to be project-based and aims to develop twenty-first-century skills—communication, collaboration, critical thinking, and creativity. Students are viewed as co-creators and entrepreneurs who are being prepared for jobs that are yet to exist. Technology is free and easily accessible, with creativity plans replacing traditional lesson plans as teaching and learning become personalized because learning now occurs both in and out of the classroom. In this way, Education 4.0 is viewed as implementing transhumanism, where the focus is on improving human beings by calling on "people to take control of the evolutionary process through technological knowledge in order to free the human species from its biological constraints" (Huk, 2021, p. 41). At the center of Education 4.0 is the continuous improvement of AI by integrating social, cultural, and educational activities into technology to enable real-time monitoring and diagnosis. Therefore, we can think of Education 4.0 as being linked to Industry 4.0 through the use of "The Internet of Things, Cloud Computing, Big Data Analytics, Autonomous Process Organisation, Augmented Reality, Horizontal and Vertical Integration, and Advanced Robots and *Co-robots*" (Huk, 2021, p. 42).

Education 4.0 is an innovative approach to education that integrates advanced technologies like online assessments, robotics, AI, Big Data, virtual reality, and augmented reality and is designed to develop twenty-first-century skills. This educational paradigm seeks to address the needs and opportunities of the Fourth Industrial Revolution. While machines cannot replicate human qualities like judgment, empathy, persuasion, collaboration, communication, flexibility, adaptability, and resilience, recent discussions have explored the potential of AI in education. The COVID-19 pandemic has accelerated the transition to online teaching and learning, leading to a gold rush in education with several AI-branded products competing for a place in education. Intelligent assistants performing administrative responsibilities for teachers, such as attendance tracking, lesson plans, and class activity development, are seen as efficient strategies to free up teachers' instructional activities. Before the COVID-19 pandemic, the incorporation of technology in education was progressing slowly. However, the convergence of Education 4.0 and AI presents an unprecedented opportunity for AI to transform educational systems. This is due to the rapid advancement of technology, which has surpassed previous assessments. In today's data-driven era, virtual reality, augmented reality, and chatbots (software programs that simulate human conversation) have become commonplace, prompting Education 4.0 to prioritize enhancing students' learning experiences.

In the digital society era[2] of "capitalism 4.0" (Kaletsky, 2011)[3] and the Fourth Industrial Revolution, the scientific-technological-industrial complex has altered educational systems. In this way, globalization 4.0, industry 4.0, and capitalism 4.0 are all inextricably linked to each other in that while their processes are different (adaptive social systems that mutate and evolve in response to a changing environment), their outcomes are the same; that is transforming the financial neoliberal landscape in response to periods of social and economic upheaval. In this way, the educational intelligent economy is undergirded by globalization 4.0, industry 4.0, and capitalism 4.0, which have brought about an age of disruption, given that they are all driven by technological developments. All human activities cannot be separated from using IT. AI is considered the "new oil" that must be extracted and refined from data. In the digital era, IT has emerged as both a tool and mechanism that aids in human life activities. Rapid

---

[2] The digital society/economy is defined by the movement away from the post-Internet society. It is driven by an exceptional socioeconomic environment in which electronic communications prevail, and cybersecurity has increased importance.

[3] Capitalism should be viewed as an evolutionary system, whose economic rules and political institutions are subject to profound changes based on social and economic crisis. In the early stages of capitalism, commonly referred to as capitalism 1.0, politics and economics were perceived as distinct domains. In contrast, capitalism 2.0, which emerged during the 1930s–1970s, was marked by a belief in a strong government and interventionist policies. Conversely, capitalism 3.0 (1970s–2000s) was defined by the Thatcher-Reagan revolution and their reversed of the structural deterioration of Anglo-Saxon capitalism and a focus on the growing emphasis on market fundamentalism (Gamble, 2012). In the post-2008/2009 global financial crisis has given rise to capitalism 4.0.

advances in Big Data (which combines data-management technologies) and AI technologies, IoT, and other ICT have profoundly impacted all aspects of human society. AI (which includes many automation capabilities) is developing rapidly, so it is almost impossible to keep up with new technological advances. As such, we distinguish between education technology (EdTech) at large and AI in education (AIED). EdTech is the practice and theory of combining learning approaches to education with ICT tools to enhance learning. While AI-powered solutions have been around for a while in the EdTech field, the industry has been sluggish to embrace them and "what distinguished the field of AIED from educational technology was the systems' ability to make instructional decisions, so that different learners could be offered different exercises, or a different order of exercises, or could be asked to present the concepts to be learned in a manner tailored to their level" (Poellhuber et al., 2024, p. 152). AIED techniques (e.g., natural language processing, artificial neural networks, machine learning, deep learning, and genetic algorithms) have led to the building of "faster classrooms" (Roll & Wylie, 2016) as these focused on combining traditional modes of education with instructional design, technological development, and education research to create intelligent tutoring systems, teaching robots, learning analytics dashboards, adaptive learning systems, human-computer interactions, dynamic immersive simulations, and adaptive courseware (Chen et al., 2020; Ouyang & Jiao, 2021). These AI-based systems utilize various data of learners, such as their scores, clicks, behaviors, and even emotions, to evaluate their skills, provide feedback, recommend strategies, and perform other functions. AIED uses technology to connect educational and learning theory to instructional design and technological development by using technologies to facilitate teaching, learning, or decision-making. AI-based software also plays a role in decision-making processes related to college admission, financial aid, course placement, enrollment, and other educational activities beyond the realm of learning. We argue that AI-based approaches in education are favored to produce more efficient, adaptable, comprehensive, customized, and engaging learning environments using pedagogical, learning domain, and learner models. We agree with Roll and Wylie (2016) that "AIED research should strike a balance between evolution (refining existing frameworks) and revolution (thinking more broadly and boldly about the role of ILEs)" (p. 583).

In the wake of the global COVID-19 pandemic, companies have been scrambling to gain a foothold in the lucrative educational technology market by proposing new AIED products. Major companies (such as Amazon, Alphabet [Google], and Meta [Facebook]) and educational technology firms (such as Knewton, Carnegie Learning, and Global Understanding XPrize) contribute significantly to developing AIED products. In 2022, investment in generative AI was just over USD 1 billion. In 2023, Microsoft poured USD 10 billion into Open AI, the company behind ChatGPT. With the expansion of digital devices and university apps, the acceptance of e-learning and the rise of MOOCs, access to large amounts of data, and outstanding advances in adjacent disciplines such as machine learning and data mining, the Global Market Insights ([2021] www.gminsights.com) argues that the AIED market has accelerated its growth speed by 40 percent per year and is projected to exceed USD 20 billion in 2027, up from USD 400 million in 2017.

COMPARATIVE AND INTERNATIONAL EDUCATION AS A COLONIAL PROJECT

Before we consider AI's impact on CIE, we must first recognize that the field of CIE is a colonial project currently seeking to decolonize itself within the context of the Fourth Industrial Revolution, which blends the physical with the cyber-physical, and is driven by the educational intelligent economy, which is premised on the exponential production of digital data to measure, analyze, and predict educational performance in comparative perspective (Salajan & jules, 2020). While some have traced the earliest roots of comparative education to the ancient Romans, Greeks, and Persians (see Brickman, 1966), the origins of the field as a *professional field of study* trace its origins back to Marc-Antoine Jullien's work, *Esquisse d'un ouvrage sur l'éducation comparée*, in 1817. Since then, scholars have treated the field's history as one characterized by evolutionary phases (the traveler's tales, educational borrowing, international education co-operation and educational exchanges, studying the national character of the educational system, and scientific perspective) that follow and ultimately eclipse earlier stages (Epstein, 2008). This focus on evolutionary stages has blinded CIE to reckon with its colonial past. In fact, many scholars writing about the field's history tend to focus on the evolution of positivism within the field, or cite Michael Sadler's relativism, which is codified in his famous speech given in Guildford, England, in 1900, where he states that

In studying foreign systems of Education we should not forget that the things outside the schools matter even more than the things inside the schools, and govern and interpret the things inside. We cannot wander at pleasure among the educational systems of the world, like a child strolling through a garden, and pick off a flower from one bush and some leaves from another, and then expect that if we stick what we have gathered into the soil at home, we shall have a living plant. A national system of Education is a living thing, the outcome of forgotten struggles and difficulties and of battles long ago. It has in it some of the secret workings of national life. It reflects, while seeking to remedy, the failings of national character. (cited in Hans, 1958, p. 3)

While the foundation of the field in the nineteenth century embraced "two opposing epistemologies—the universalism of positivism and the particularism of relativism" (Epstein, 2008, p. 377), it never fully addressed coloniality and the impact of colonialism in establishing the field. Instead, the field was polarized between positivists such as Harold Noah, Max Eckstein, Arnold Anderson, and Philip Foster, who argued that the role of comparison should be to look at differences and similarities between units amenable to analysis and relativists such as Vernon Mallinson and Edmund King who claimed that comparison came from gaining deep insight—*Verstehen*—through the understanding of the "forces" and "factors" of the educational structures and functions of a different nation or group (Epstein, 2008). The 1930s through to the 1960s saw the rise of historical functionalism—examining education as "interrelated with other social and political institutions; and it can best be understood if examined in its social context" (Kazamias & Massialas, 1982, p. 309)—and even further neglected theorizing colonialism. CIE does not discuss how the field has exoticized the "Other" in the study of so-called backward peoples and cultures and the quest to expand and impose universalized notions of Western knowledges. As the field has sought to study other people and cultures in comparative perspectives, its ontology, epistemology, axiology, and onto-epistemology are often grounded in Western (usually colonial) ways of knowing that inform Eurocentric settler colonial notions of the "Other." It was not so long ago that some comparativists in our field echoed sentiments such as those conveyed by Brickman (1950), who suggested that the aim of comparative education, as a subset of international education, included the "rehabilitation of backward cultural areas" (p. 617) or propagated notions similar to that of Paul Monroe, the first director of Teacher College's International Institute, who expressed an interest "in the development of retarded cultures of backward peoples through the instrument of education" (as cited in Goodenow & Cowen, 1986, p. 277; see also Bu, 1997; Takayama et al., 2017).

In the early days of the field, communication was limited to correspondence, reports, and publications with limited circulation. However, with the advent of the typewriter, textbooks became the primary mode of communication during the modern scientific period. The first Western textbook, *Comparative Education*, was published in London in 1918 and edited by Peter Sandiford of the University of Toronto, Canada. Isaac Kandel's *Studies in Comparative Education* in 1933 was another significant milestone. Chinese scholars also made their mark with four books entitled *Comparative Education*, published between 1929 and 1934. Recent neoliberal educational epochs (defined by the implementation of structural adjustment programs, the marketization of education, the hollowing out of the state through Thatcherism[4] and Reaganomics,[5] and the rise of the Washington Consensus and post-Washington Consensus) have created the era of educational datafication (our data footprints) defined by the vast amounts of data that are produced and consumed in educational settings. This has been exacerbated by the movement from government to governance. As the field experiences a vibrancy, increased attention has been placed on the role of ICT in impacting its traditions with a growing recognition that, since its inception, the field of CIE has undergone significant technological advancements.

Despite these milestones, the field of CIE has remained a colonial project that has never really dealt with its colonial past. It is essential to consider the consequences of using AI, as humans train it and can be prone to biases or "hallucinations." AI may provide inaccurate, biased, or unintended information when trained on a limited data set. Therefore, it is crucial to be aware of AI's limitations and ensure it is used responsibly. The issue at hand is of great concern as it was only after the widespread dissemination of the Black Lives Matter Movement and the global movements for racial justice in education that CIE began to take into account the macro-level structural forces of settler colonialism, state violence,

---

[4] Thatcherism attempts to promote low inflation, the small state, and free markets through tight control of the money supply, privatization, and constraints on the labor movement.

[5] Reaganomics had four simple principles: lower marginal tax rates, less regulation, restrained government spending, and noninflationary monetary policy.

racial capitalism, anti-Blackness, and erasure of Indigenous people. Cowen (2000) suggests that delving into the histories of CIE provides a sense of comfort by reinforcing a shared identity, offering fodder for conversations, and serving as a convenient response to inquiries like "What is comparative education?" (334). However, an examination of the present-day field reveals a perpetuation of an education history steeped in colonialist, racist, sexist, and ableist theories. Hayhoe and Mundy (2008) describe this phenomenon as the "darker side of comparative education" (p. 5), marked by the utilization of comparative research in shaping and reforming colonial education for purposes such as fostering democratization, global peace, educational collaboration, and international understanding. This stems from the fact that the origins of the field are often tracked back to the "pioneers," all of whom are white males. Figures like Marc-Antoine Jullien are hailed as the founding fathers, followed by well-known names predominantly from the Global North—Michael Sadler, Isaac Kandel, Nicholas Hans, Friedrich Schneider, Edmund King, Brian Holmes, William Brickman, George Bereday, Harold Noah, and Max Eckstein. In short, the field's history has been framed "around individual intellectual brilliance and collective progress, providing a comfortable framework for wrapping our identities and feeling secure about who we are and what we stand for" (Takayama, 2018, p. 460).

DECOLONIZING ARTIFICIAL INTELLIGENCE IN THE ERA OF THE ANTHROPOCENE: PERSPECTIVES ON DIGITAL HUMANISM

We must remember that there are two types of AI. Strong machines that deploy deep learning using neural networks to think creatively and develop self-consciousness. Weak AI implies machines' ability to take on human activities, such as automation. As such, to begin to think about decolonizing AI and its implication for CIE, we must begin by framing AI in the current context of the Anthropocene. Given that the Anthropocene has become defined by human industrial activities on the planet through human-led modification and exploitation of the environment, the proliferation of AI will be a significant driver of change in shaping our present and future trajectory in relation to the Anthropocene. In essence, human-environment interactions have created the "technosphere," defined as "the human-created fabric of industrial technologies, infrastructures, energy flows, and social institutions that increasingly interact with and function at a level equivalent to that of other Earth system spheres" (Creutzig et al., 2022, p. 482). Thus, AI-powered systems, while not a silver bullet, can play a significant role in helping us undo the damage caused to the environment. In fact, a recent review of the evidence shows that AI can act as an enabler in helping us achieve 134 targets (79 percent) across all Sustainable Development Goals (SDGs) (Vinuesa et al., 2020). However, as Peyron (2021) reminds us, AI "transforms society, economic and political relations: if our action is aimed at reducing, limiting, mitigating or reversing the misuse of human presence in the ecosystem, the so-called Anthropocene, the question that arises is whether to increase the Anthropocene is the correct way to reduce the Anthropocene" (para 8, authors translation). Thus, the current era of the Fourth Industrial Revolution is different from other epochs in that we live in a "data driven, algorithm-mediated economy of the twenty-first century" (The Economist, 2019, p. 1) centered on AI-powered industrialism. Technology originates from knowledge, which, in turn, stems from information. In the contemporary landscape, technological advancements have notably heightened the accessibility of information to an unprecedented degree. Scientists are asking: "Are we humans defining technology or is technology defining us?" (Lee, 2020). Thus, technology is a top-down intelligent design or what Lee (2020) calls "digital creationism," in that AI turns information into a tsunami, where every image, text, and sound are turned into more information that floods the human brain. In this way, we need to think of digital technologies as *co*evolving with humans in that as humans change, technology changes, and therefore, we need to think about how we can regulate technology and bring about mass access to it.

As such, to understand the impact of digital transformation on the field of CIE, one must observe the field through a decolonial lens. Such a lens helps us understand the power patterns in the social and technological fabric that we have inherited along with other patterns of inequality or what can be called "coloniality" (Mignolo & Walsh, 2018; Quijano, 2016). This is because the structure of the global digital economy is extremely centralized and concentrated in the Global North. For example, when the Internet, the backbone of the digital society, was developed, its inventor, Tim Berners-Lee, argued that it would be "an open platform that would allow everyone, everywhere, to share information, access opportunities and collaborate across geographic and cultural boundaries" (Berners-Lee,

2017, para 1). However, at the time of writing, while the digital divide has been decreasing, some three billion people in the world are unconnected from the digital society. We cannot let this happen with AI.

We have not considered "the complex genealogy of intelligence: whose conception of intelligence is modelled within technology or how the idea has been put to work in dividing people between the desired and the undesirable" (Adams, 2021, pp. 176–77). As such, we must consider that AI is built upon knowledge advanced by imperial powers as a way to control and contain colonized populations. Scholars such as Mbembe (2017) contend that in our global society, we are currently encountering the third shift in the arrangement of race and blackness—with the first being colonization and slavery; the second is the decolonization process; and the third is the advent of technological innovation that represents the current phase of high modernity. Similarly, Quijano (2017) notes that we are at a critical turning point around the global coloniality of power, with "the manipulation and control of technological resources of communication and of transportation in order to impose the technocratization/instrumentalization of Coloniality/modernity" (p. 364). As AI and digital technologies commodify human experience, this implies that new digital technologies are intensifying and amplifying pre-existing inequalities, particularly those rooted in race, ethnicity, and national origin. Coupled with the extraction and exploitation of personal data and data systems, we are seeing the rise of "data colonialism" or "digital colonialism," as well as the advent of "AI nationalism" (Hogarth, 2018), which is creating new divides and exacerbating old divides between the Global North and South.

Technological coloniality, "tech evangelism" (Birhane, 2019), and "algorithmic colonization" (Birhane, 2019; Murphy & Largacha-Martínez 2022) can be observed in many sectors of society as "data, knowledge, expertise, and high-performance infrastructures are kept and mined by an increasingly smaller number of transnational corporations, using highly advanced digital technologies for value extraction and profit" (Bon et al., 2022, p. 63). This is because algorithms (which have been cloaked in logic and mathematics) have been given a seigniorial status, consistent with the myth of Western rationality around universals such as ideas, God, and natural laws. Because algorithms "are treated as idealized formulas that operate in ways that supersede human capabilities" (Murphy & Largacha-Martínez, 2022, p. 102), AI reinforces the Cartesian dualist correlation between science and reason and performs a similar role in the colonialism of knowledge. Several scholars have argued that coloniality can be observed in AI algorithms that humans train and were once taught as value-free but are now viewed as having biases. From a colonial perspective, we can see AI as a nontransparent socio-technical system built upon knowledge production with its own ontologies and epistemologies. Central to this system are the relations between power and knowledge, which accounts for how power is maintained through socially accepted forms of knowledge, such as scientific understandings, ethical norms, and codes, or what is commonly accepted as "truth" (Foucault, 1978). Thus, if assumptions and structures behind these kinds of knowledges "are left unexamined and unchallenged, [they] may result in continued forms of bias against historically oppressed bodies" (Nemorin et al., 2023, p. 39). Thus, for AI to bring about the positive benefits in education that are often touted, such as spurring digital transformation, innovation, and growth, reducing access barriers, automating management processes, optimizing learning outcomes, and closing the digital skills gap, it must be decolonized. Only through the decolonization of AI can we reap the benefits (economic, political, and social) that AI has to offer. As the EU High-Level Expert Group on AI (2019) argues, "AI is not an end in itself, but rather a promising means to increase human flourishing, thereby enhancing individual and societal well-being and the common good, as well as bringing progress and innovation" (p. 4). Moreover, AI has become central to the discourse on restructuring school governance, pedagogy, and learning methodologies to align with the acquisition of OECD twenty-first-century skills deemed essential for active participation in broader societal contexts. However, the question beckons: *how do we decolonize AI?*

To think about decolonizing AI is to accept that coloniality is the Janus-face of capitalism and modernity in that without slavery and colonialism, there would be no modernity and capitalism (Adams, 2021; Mignolo, 2009). Thus, the end of modernity/capitalism dynamics is "the ultimate decolonial horizon" (Mignolo & Walsh, 2018, p. 4). Similarly, for Quijano (2000), *de*coloniality is constitutive of the inverse force of colonialism in that it seeks to unravel the "coloniality of power" by

> Identifying the signifying practices (the ways in which authoritative meaning is (re)produced) of the coloniality of power—the epistemic, cultural, political and economic apparatuses through which the oppression of (sub)alterity is constituted and maintained, and Eurocentric ways of knowing, being, and thought are reproduced as superior and singularly legitimate—is globally critical to the decolonial project. (Adams, 2021, p. 180)

Therefore, the decolonization of AI is warranted in that AI, in its current form, carries the vestiges of colonialism that continue to shape experiences of subjugation and oppression along racial lines because of the existing "latent and overt systems and structures of racial (and gendered) bifurcation that mark, divide, categorize, classify, and hierarchize individuals according to social norms of un/desirability" (Adams, 2021, p. 180). If we are to think of the pathology of race as undergoing a series of three significant arrangements—with the first occurring in the seventeenth century around the fabrication of the myth of European superiority, the second in the nineteenth and early twentieth century, which categorized race as scientific discourse, and the third from the 1930s onwards where race became viewed as a social and ideological construct—then decoloniality represents the final stage of these arrangements (Soudien, n.d as cited in Adams, 2021). Thus, to decolonize AI is to decolonize language since "language carries culture, and culture carries, particularly through orature and literature, the entire body of values by which we come to perceive ourselves and our place in the world" (Wa Thiong'o, 1986, p. 16). Therefore, a more human form of AI that is decolonized would recognize different value systems and provide alternatives to the Eurocentric ways of knowing and being that do not "recenters whiteness, ... resettles theory, ... extends innocence to the settler, [and] entertains a settler future" (Tuck & Yang, 2012, p. 3). This implies "pluriversal" thinking, which is grounded on the notion that other worldviews (outside of Western/Eurocentric ideas of modernity) are possible and interconnected, especially Indigenous relational worldviews (Mignolo & Walsh, 2018). Such a move requires a reexamination of Western forms of knowledge and a radical pluralization of what it means to be human (Césaire, 2001; Wynter, 2003). In other words, while decoloniality is historically situated, it " ... is the strategy, a situated and responsive strategy to the particular form of power over that coloniality impressed upon the world, and not an autonomous and separable concept that can be put to work as a strategy for something else" (Adams, 2021, p. 181).

Thus, decolonial datafication begins by having a common framework that promotes responsible AI in an ethically and legally sound environment. This involves detaching "ourselves from the semiolinguistic limits placed on our own imaginative realm due to the afterlife of colonialism and racial slavery" (Alagraa, 2018, p. 165). Such a move draws upon Wynter's (2003) work, which is situated within a tradition of anticolonial and decolonial thought and argues that "man" is overrepresented and such "overrepresentation is therefore a problem of the overrepresentation of the West's genre-specific truths, through colonial occupation and racial slavery (among other methods)" (Alagraa, 2018, p. 165). Since notions of AI are entrenched in epistemological legacies of colonialism, which are a consequence of the processes of coloniality, decolonizing it involves a type of "epistemic disobedience" (Mignolo, 2009), which calls for thoughtful attention to the silences of Western epistemologies, excavating those silences, and affirms the epistemic rights of the margins. In this way, coloniality, which moves beyond the historical project of colonialism and imperialism, recognizes the efficacy of colonial power relations and its lasting marks on AI. Decoloniality, then, requires delinking from coloniality and modernity. To decolonize, therefore, is to break the rapture and address the conceptual gaps left by colonialism. This involves breaking away from the construction of Western technological philosophy. Such a reparative reading compels new ways of imagining AI's benefits to education by re-embodying and relocating Western thought to unmask the limited situation of modern epistemologies and their linkages with coloniality. This means delinking oneself from these knowledge systems and reimagining other futures where AI is open to all.

To decolonize AI in the context of CIE is to question how power works through technological systems, such as ethics. Ethics, which stems from Eurocentrism and is viewed as universally applicable, has become the "language du jour" (Ulnicane et al., 2021) in AI discourse and is thus viewed as one of the first things that can be decolonized, given how ethics has been used to rationalize colonial practices (Adams, 2021). We often forget that ethics is a value-laden concept that reaffirms white superiority, given that it has historically functioned as a racialized dividing practice to differentiate between peoples. For example, if AI ethics are not appropriately decolonized, it runs the risk of reproducing the very logics of race that colonialism instituted. Adams (2021), in an analysis of AI ethics standards, found that none of the 81 standards published by the Global Landscape of AI Ethics Guidelines were from the Global South. Similarly, Ulnicane et al. (2021) note that by 2020, at least 50 countries have developed or are in the process of developing national AI strategies. Nevertheless, as Mohamed et al. (2020) point out, these national policies were all in the Global North. As such, with the rise of "ethics dumping" (Mohamed et al., 2020), where inadequate data protection is exploited across the laboratories of the Global South, we see a pattern where the South is portrayed as "pre-ethical" (Mbembe, 2017) and in need of saving by Western knowledge and rationality (for example, the development of algorithmic systems for use in the US and UK by Cambridge Analytica and the beta-testing them in Nigeria and Kenya). This is problematic because AI systems sort personal data according to socially ascribed normative markers (which are often racialized, gendered, implicitly biased, etc.), and thus, it can re-inscribe the divisive logics of race and racialized hierarchies, imperial power patterns, and enforce social stratification and biases. In working this way, AI systems risk further marginalizing those who do not fit into its archetype and thus "de-equalizing" (Quijano, 2017) people.

Conclusion

In this chapter, we have discussed the implication of the TIC on Education 4.0. We also discussed that the field of CIE is a colonial project that is seeking to decolonize and that this has consequences for the Anthropocene. We conclude here by arguing that given CIE's foundations, we must employ a decolonial praxis toward how AI is employed in the field. This is because decolonization is about undoing the legacy of colonialism, using different knowledge sources, acknowledging diverse perspectives, and empowering marginalized communities and vulnerable populations. When it comes to AI, there is a risk of perpetuating the violent imposition of social hierarchies and biases if the data used to train models reflects historical inequalities. This is because the historical data we rely on reflects societal prejudice, which can inadvertently reinforce and perpetuate discriminatory patterns when used to train AI models. Thus, ensuring that the data used to train models and the data teams are diverse and focused on fairness is necessary to decolonize AI within the context of CIE. Many have argued that for AI to fulfil its true potential, it must be decolonized so that it does not become a vector of a new kind of colonial harm. Education is one area in which AI can bring significant benefits to humanity as it has the potential to increase access, accountability, and decrease inequality, but educational systems are fundamentally compromised by European colonial erasure of communities, and this has implications for contemporary education AI. Zimeta (2023) maintains that "most developments in AI are driven by the kinds of extractive business models that drove European colonization, with many of the same harmful impacts" (para. 13), which implies that AI has the potential to "exploit the global poor, introduce digital apartheid and create new forms of political disenfranchisement and coercive surveillance around the world" (para. 15).

Decolonization is not only about establishing a more human-centered version, but it should seek to address the injustices of colonialism by rapturing the epistemological and teleological assumptions that AI is built upon by historically reading the formation and appropriation of colonial regimes given that coloniality continues to operate in technologies and imaginaries associated with AI. Across the rest of this book, we pick up on these themes and employ a decolonized and social justice lens to discuss ways in which CIE can handle the onslaught and proliferation of new AI-related developments.

CHAPTER 3

The Emergence and Progression of AI in Comparative and International Education

Abstract The history and evolution of AI reveal many twists and turns, but a consistent thread in the story of AI is the humans' fascination to (re) create intelligence in their own image. Ancient myths of intelligent machines serving humans in their mundane or ambitious undertakings stored in human imagination and consciousness throughout the centuries have infused modern attempts at creating machines capable of displaying intelligence comparable with those of their human creators. In this process, humans have brought to bear cognitive science, psychology, philosophy, technology, and computer science to tinker at the intersection between human biology and neuroscience with computer programming, software and hardware to breathe life into the enduring, yet still elusive, captivation with the AI machine. This history is replete with cautionary tales for education in general and comparative and international education in particular, as AI systems and platforms are attempting to (re)shape the educational landscape with important consequences for the foreseeable future.

Keywords History • Emergence • Evolution • Mythos • Reality

INTRODUCTION

The dawn of AI is as nebulous as the term itself is elusive, and experts hardly agree on or can determine the precise point of departure for the emergence of this intriguing, appealing, unavoidable, and increasingly pervasive phenomenon. Part of the difficulty in establishing with any measure of confidence and precision the genesis of AI in the current era rests in the extent to which the concept is conflated with the development of modern technologies, from mechanical apparatuses to sophisticated digital programs and applications or whether it is treated as an encompassing notion tied to elements of human cognition, sentience, consciousness, and self-awareness that it is envisioned to emulate or replicate. As such, in some interpretations of the purpose and meaning of AI, "the long-term goal of artificial intelligence is to create a thinking machine that is intelligent, has consciousness, has the ability to learn, has free will and is ethical" (O'Reagan, 2016, p. 250). Further, the form in which non-biological intelligence manifests itself and can operate autonomously from human control or commands in such a way that it evolves on its own has been a question that has puzzled philosophers, scientists, ethicists, legal experts, and educationists alike for centuries. Neither has the question of what precisely constitutes intelligence in human and artificial/machine form been definitively settled. Moreover, philosophical questions related to the nature of mind and consciousness and their relationship to the (human) body have been contemplated since antiquity, especially by Greek thinkers, continuing through the ages until modern times.

Notably, during the Enlightenment, the seminal mathematician and philosopher René Descartes laid the foundation of rationalist thought, particularly examining the duality between mind and body. He argued that the mind was the only reliable source of information about an individual's existence since thinking whether one exists or not is clearly a judgment indicating a person exists. In turn, following this logic, a person cannot assume one's body exists, as the body is perceived through one's senses, which, in his view, are unreliable and, therefore, not to be trusted as an indication of the actual existence of one's body. His deductive logic led him to conclude that he was a thinking thing, the essence of which was captured in his famous dictum, *cogito ergo sum* (I think, therefore I am), cemented in the Cartesian dualism which professes that humans possess an acute conception of self as distinct from the body. These precepts influence the rise and development of artificial forms of intelligence, particularly in their embodied manifestations, as any AI machine or system that aspires to take on human characteristics would have to exhibit the self-awareness and conception of personhood inherent in Cartesian dualism. The open question for the future of AI is whether an artificially intelligent apparatus or system can be endowed with the equivalent of human consciousness and self-awareness. That is, could further refinements to software, coding, and programming languages give rise to an artificial mind that thinks of itself in Cartesian terms and considers the hardware in which it resides and manipulates the physical representation of its body?

Operationalizing the theory of pluriversality, this chapter will explore the question of whether and how AI approaches the abilities of human intelligence. Interconnected stories told through film, literature, and scientific innovation will be used to trace the origins of AI and outline its emergence into popular culture. Beginning in the era of ancient mythology, through critical developmental milestones of modern AI technologies, we will argue that while AI is associated with computers, it is necessary to see it as much broader than just machinery embedded in computers, sensors, and other emerging technologies. Instead, in this chapter, we will consider the rise of AI as both a tool and a concept (Chassignol et al., 2018). We will also overview some of the contemporary typologies of AI. Finally, this chapter will conclude by returning to applications of contemporary AI to CIE.

From Mythos to Reality

The creation of AI or artificial life mimicking human qualities has been circulating in the realm of imagination, arguably from antiquity. The fascination with the prospect of developing intelligent machines, albeit controlled by and subservient to humans whose needs they inevitably have to meet, can be traced as far back as Homer's mythical epic poems *Iliad* and *Odyssey*. In these stories, Homer imagines and describes intelligent or thinking machines, a sort of mechanical assistants of various levels of operativity and intelligence, ranging from the relatively benign to the highly sophisticated. For example, devices such as the self-opening gates that automatically open before the Homeric gods are in close proximity, the bellows that move autonomously in Hephaestus's smelting shop, in which he produces self-propelled tripods, exhibit a level of lower-mode intelligence meant to serve the commands of its human owner, an early manifestation of what today is termed as "weak AI" (also known as "narrow AI" due to the relatively limited, rudimentary, highly specialized, and repetitive scope of the function being performed by this type of specific intelligence). Similarly, the self-navigational ships, though endowed with higher information processing powers, as in their ability to plot a route based on coordinates stemming from human verbal input, then navigate that route independent of human intervention, represent a form of upgraded AI, though still not in its highest manifestation in the Homeric poems. That privilege is reserved for a form of embodied AI endowed with thinking capacities closely mimicking those of the human intellect, as these "anthropomorphic automata possess both mind and the power of intelligent thought" (Liveley & Thomas, 2020, p. 37). Such anthropomorphic intelligent machines, able to interact with their human creators, mainly for the benefit of the latter, represent the highest form of AI contemplated in ancient mythology, a sort of original or primordial AI story that has been ingrained in human memory, giving rise to the modern concept of "strong AI" (also labeled "general AI" and characterized by a wide range of interactive functions that involve high computational power, increased processing capacity, autonomous decision-making, self-learning abilities and replicating cognitive operations typically attributed to humans, therefore constituting a form of general intelligence). As Liveley and Thomas (2020) put it,

> Homer's tales of intelligent machines are stored as part of the AI data set in the human 'memory bank' that constitutes the classical myth kitty. Homer's intelligent machines are programmed into our cultural *episteme* (understanding/knowledge) and *doxa* (opinion/belief) about AI. Homer's AI narratives, it seems, have been placed into modern human *phrenes*, into the *noos* of the Western cultural imaginary. (p. 44)

Another ancient myth contemplating the creation of artificial life endowed with human-like intelligence is the story of Pygmalion in Ovid's (Publius Ovidius Naso) fifteen-book narrative poem Metamorphoses. In one of the books of this first century AD mythological epic poem, Ovid recounts the life of a Cypriot king and sculptor, Pygmalion, who is repulsed by women because he views them as compromised, given some women's participation in the sex trade. Having stigmatized an entire gender as tainted or impure for the behavior of a few, Pygmalion takes a vow of celibacy and channels his creative energy and imagination into a sculpture of what, in his conception, constitutes the perfect woman. He subsequently falls in love with his inanimate creation and prays to the goddess Aphrodite to grant him a wife in the likeness of his statue. The goddess eventually hears the king's prayers and decides to breathe life into the sculpture, thus making her human. Pygmalion wastes no time in acting on his feelings and, after naming her now alive companion Galatea (for the whiteness of her skin, itself derived from the white marble from which it was sculpted), he makes her his wife and produces an offspring, a daughter named Paphos. Although Galatea's creation myth suggests a culmination of one man's wish to transform an inanimate likeness of a human into a living, breathing, and thinking human being, it is unclear from the narrative whether Galatea is capable of interacting "intelligently" with Pygmalion or other humans, therefore leaving only the assumption that, in the process of becoming human, Aphrodite endowed her with some level of human-like or human intelligence. Notwithstanding the inscrutability of Galatea's intelligence as portrayed in the story, her metamorphosis from an inanimate, artificially created object of one man's affection into a living and presumably feeling, thinking, and talking human being represents a poignant example of the human fascination with creating artificial life that has endured through the ages, and implicitly or explicitly continues to drive the human desire to develop artificial intelligence.

Further examples of the continued fixation with AI in human consciousness are the literary and cinematographic works of science fiction that begin with nineteenth century novels to contemporary movies or television series that continue to fuel human imagination around, but also inform tangible or practical manifestations of AI. The prolific author Jules Verne can be credited with an extensive collection of novels that present futuristic scenarios in which humans employ sophisticated machinery apparently endowed with some level of intelligence that enable their operators to achieve fantastical goals or aims, such as reaching the Moon, sailing around the world, or submersing thousands of miles under the seas or oceans. It was only a matter of time before Verne's ideas took on concrete shape in the twentieth century when computational power, the pervasive and underlying form of AI in modern innovations, made it possible for humankind to set foot on the Moon, plunge to incredible ocean depths, and master air travel by way of various flying apparatuses. Stepping into the twentieth century, another seminal science fiction author would arguably lay the foundations of modern interpretations and manifestations of AI, most closely linked with the field of robotics. In his groundbreaking short story, *Runaround,* Issac Asimov tells the futuristic story of the development of an anthropomorphic robot named "Speedy" who helps its human companions on missions related to selenium mining on Mercury. In this novel, Asimov formulates the *Three Laws of Robotics*, which postulate that:

1. A robot may not injure a human being or, through inaction, allow a human being to come to harm.
2. A robot must obey the orders given it by human beings except where such orders would conflict with the First Law.
3. A robot must protect its own existence as long as such protection does not conflict with the First or Second Laws. (Asimov, 1950, p. 40)

Although not necessarily or clearly invoked today, these laws serve at a conceptual, abstract level, not in a real, practical sense, as a form of ethical principles in developing AI technologies, particularly in robotics. They especially inform, albeit implicitly, about the development of anthropomorphic robots, intended for the time being as technical, office, or personal assistants to humans. However, such mechanisms built in the image of their creators, though impressive as manifestations of human ingenuity in AI development, are currently mainly in prototype stages, still far from being fully autonomous and falling short of emulating all the innate characteristics of human nature. Nevertheless, once cinematography made it possible for humans to express their imagination through moving pictures, it did not take long for the fascination with anthropomorphic AI to take center stage on the silver screen. One of the earliest productions envisioning the cohabitation of humans with AI is Stanley Kubrick's *2001: A Space Odyssey* (1968), a poignant visual representation of Arthur C. Clarke's Space Odyssey novel series, depicting both the marvels and dangers of AI. In the story, a crew of astronauts on a mission to Alpha Centauri navigate through the void of the cosmos in a state-of-the-art spaceship, in which control is relinquished almost entirely to HAL 9000, a highly advanced on-board computer that monitors and commands virtually all the ship's functions. Although not embodied as an anthropomorphic character, HAL can perform highly sophisticated computational functions necessary for interstellar travel and emulate human communication, intelligence, sentience, and even emotions. After a series of malfunctions apparently motivated by a sense of self-preservation and detected by the two protagonist astronauts, HAL eventually becomes fearful of its fate when it learns that the two crew members plan on taking it offline. In the end, HAL is disabled by Dave, the only surviving astronaut on board, in an expression of the control humans still exercise over AI.

Since then, similar fictional stories of AI, particularly envisioned in its embodied form as anthropomorphic characters, have populated the cinematographic space. An illustrative example of the ultimate AI embodiment is Lieutenant Commander Data, the second officer and chief operations officer aboard the starship Enterprise in the *Star Trek: The Next Generation* (1987–1994) television series. As a sentient, self-aware, fully autonomous, and functional android, Mr. Data is virtually indistinguishable from his human colleagues and represents, at least in the realm of imagination, the culmination of humans' endeavor to create artificial life and intelligence in their image. He not only performs the functions and duties inherent to his rank, exhibiting immense computational power in the process, but he deliberately seeks to act like a human endowed with all of the emotions and social abilities of a human being, a desire he expresses that is unclear whether it is a consequence of his programming or his evolution as a sentient being and his conscientious aspiration actually to become human. Another science-fiction series popularizing anthropomorphic AI beings is *Star Wars*, in which viewers are exposed to a wide range of androids and AI machines that serve benign or malign intentions. It is a rich visualization of a futuristic intergalactic world rife with conflict and tension in which humanoid or mechanical AI devices serve both the benevolent and malevolent goals of their human masters. The *Terminator* series similarly portrays brilliant anthropomorphic machines sent back in time from a post-apocalyptic future to assassinate John Connor, the leader of the human resistance movement against Skynet, a superintelligent neural network that becomes self-aware and begins the process of exterminating or eliminating the human race to assert complete dominance over all earthly domains. Finally, in Steven Spielberg's seminal film *A.I. Artificial Intelligence* (2001), the viewers learn about the life of a mechatronic boy, David, built to love and fill the sense of void his parents experience when losing a child, on his journey to discover his nature and meaning in life. In following his desire to become a real boy to ingratiate his adoptive mother, David channels the imagery of Pinocchio, which likely informed at least part of the plot in Spielberg's movie. The similarity with Carlo Collodi's story is telling, as Geppetto's motivation to create a wooden sculpture of a boy, which he then wishes to come alive as a real boy, stems from his grief of losing his son. Pinocchio's and David's stories evoke and resonate with the long-standing human fascination with creating artificial life and are part of a constant thread in human consciousness. Yet, these stories also have in common the ability for humans to exert control over their creations and, as such, AI is intended to be subordinate to and serve human desires, needs, and aims, rather than endowing AI, particularly strong AI, with self-awareness and consciousness. Therein lies an ontological conundrum for AI, as humans grapple with the tension between unleashing their unbridled imagination to develop artificial life in their likeness and limiting the extent to which AI should or could become self-actualized, autonomous, and sentient beings, capable of replicating themselves and seeking legitimacy as societal actors entitled to protect their own wellbeing and to pursue their own interests.

While the arc of imagination on AI, in its broader interpretation, stretches back thousands of years, contemporary visions of AI gradually translate into real manifestations of the phenomenon. As such, a watershed moment in the modern development of AI came in the 1940s with Alan Turing's development of a computer named *The Bombe*, intended to break the Enigma code used by the Nazis during World War II, a task that had eluded some of the leading mathematicians up to that point in time. Eventually, Turing's remarkable feat became the basis for what has been referred to as Turing's test, originally called the imitation game. In his article, *Computer Machinery and Intelligence*, Turing sets the premise of the test by asking whether machines can think, then proceeds with providing a series of objections, from the theological and mathematical to the concept of consciousness, acting as arguments against the prospect that machines, in fact, do think. Although, in the end, the idea of thinking machines is partially refuted by Turing, he nonetheless refrains from discounting it in the future when he predicts that computational power would only increase, a scenario he captured as follows:

> I believe that in about fifty years' time it will be possible, to programme computers, with a storage capacity of about 10^9, to make them play the imitation game so well that an average interrogator will not have more than 70 per cent chance of making the right identification after five minutes of questioning. (Turing, 1950, p. 442)

Turing's test is now considered the standard by which the intelligence of an artificial mechanism is measured, which, summed up, postulates that "if a human is interacting with another human and a machine and unable to distinguish the machine from the human, then the machine is said to be intelligent" (Haenelein & Kaplan, 2019, p. 7). However, objections to the salience and validity of Turing's test in measuring a machine's level of intelligence soon appeared, none more prominent than John Searle's Chinese Room thought experiment. Searle contended that a machine, with all its computational power, lacked an important quality expressly reserved for humans: intentionality. In brief, in his Chinese Room example, Searle sought to demonstrate that an external human observer may perceive a machine appearing to engage in human-like communication as competent in intentionally exchanging meaningful information between two parties and is, in fact, simply adept at manipulating semiotic information that passes from input to output by virtue of established rules. Conceptually, the Chinese room experiment worked as an analogy in which a human with no knowledge of the Chinese language, isolated in a closed room, would receive through an opening a set of Chinese symbols which would have to be interpreted via a rulebook to produce a response delivered through an opening at the opposite end of the room. This would give the person on the receiving end of the message the impression that the person inside the room is fluent in Chinese when, in reality, the person inside was operating on a set of codes to manipulate the symbols received. This led Searle to question the notion of machine understanding in no uncertain terms:

> Why on earth would anyone suppose that a computer simulation of understanding actually understood anything? ...For simulation, all you need is the right input and output and a program in the middle that transforms the former into the latter. That is all the computer has for anything it does. To confuse simulation with duplication is the same mistake, whether it is pain, love, cognition, fires, or rainstorms. (as cited in Nilsson, 2009, p. 386)

Nonetheless, not long after Turing formulated his famous test, the term "artificial intelligence," in its modern conception associated with computer science and, by extension, to the use of digital computing in the development of AI, was launched in 1956 at the *Dartmouth Summer Research Project on Artificial Intelligence* (DSRPAI) hosted by the Dartmouth College in New Hampshire. As the title implies, the proposers intended to gather together a group of ten scientists for two months during the summer of 1956 to study how machines use language, how they can formulate abstract notions and concepts, and how they can engage in learning that leads to self-improvement, characteristics the authors contended were the exclusive domain of humans. Hence, they envisioned that the DSRPAI workshop would "proceed on the basis of the conjecture that every aspect of learning or any other feature of intelligence can in principle be so precisely described that a machine can be made to simulate it" (McCarthy et al., 2006, p. 12). For these reasons, the DSRPAI is considered the first scientific effort to develop artificial intelligence and McCarthy, Minsky, Rochester, and Shannon are generally credited as the "founding fathers" of the contemporary movement toward an empirical understanding and the emergence of the scientific field of AI.

A decade later, one of the first genuine attempts to implement the Turing test came in the form of *ELIZA*, a natural language processing computer program created by Joseph Weizenbaum to simulate communication between a human and a machine as closely as possible to render it to the unsuspecting user as human-to-human communication. The program replicated a psychotherapy session script, with the "patient" or the user inputting personal thoughts or questions to the "therapist" behind the screen, and the latter provided non-directional questions as output. The program was sophisticated enough to deceive some of its early users into believing they were interacting with an actual human (Weizenbaum, 1976). Not incidentally, *ELIZA*'s name harkens back to the Pygmalion story alluded to earlier in this chapter. Weizenbaum named the program after the character Eliza Doolittle from George Bernard Shaw's play *Pygmalion*, since *ELIZA* could be improved incrementally by its users, just as Eliza Doolittle could improve her speech by taking elocution lessons from Professor Higgins. More intriguingly, in echoing the parable of Pygmalion and Galatea, ELIZA draws parallels between the king-sculpture and professor-learner relationships, as a metaphor for transforming and perfecting human nature, in its subjective appearance and abilities. In that sense, ELIZA reflects the enduring power of myth in the creation and modern development of AI, blurring the lines between allegory and science. Yet, while in his article describing ELIZA, Weizenbaum (1966) recognized the potential and power represented by AI as "machines are made to behave in wondrous ways, often sufficient to dazzle even the most experienced observer," he was nonetheless keenly aware of its limitations since "once a particular program is unmasked, once its inner workings are explained in language sufficiently plain to induce understanding, its magic crumbles away" (p. 36). In his later works, Weizenbaum (1976) became an AI skeptic, seeing it as a threat to humankind because AI lacked qualities specific to humans, such as emotion, compassion, and wisdom, which are necessary to make distinctions between choices and decisions. Since humans can make judgments based on complex calculations involving emotions, Weizenbaum considered that AI should not be trusted with critical decisions directly impacting humans, as AI systems are devoid of the abovementioned human qualities, relegating decisions to purely computational algorithms.

Among other early achievements in the modern history of AI was the General Problem Solver (GPS) invented by Herbert Simon, J.C. Shaw, and Allen Newell. GPS was a computer program envisioned to function as a universal problem-solving system, which could solve relatively simple problems, such as the Towers of Hanoi—a puzzle involving three rods and a stack of six disks of decreasingly smaller diameters which have to be moved between rods within certain rules—but could not complete more complex tasks, such as solving real-world problems. Nonetheless, these achievements led to increased interest in and, consequently, funding for AI research, sparking a wave of optimism at the time about the future of AI. Enthusiasm for what lay in store for AI was so high that it led Marvin Minsky, one of the Dartmouth "founding fathers" of AI, to proclaim in a 1970 *Life Magazine* interview that,

> In from three to eight years we will have a machine with the general intelligence of an average human being. I mean a machine that will be able to read Shakespeare, grease a car, play office politics, tell a joke, have a fight. At that point the machine will begin to educate itself with fantastic speed. In a few months it will be at genius level and a few months after that its powers will be incalculable. (Life Magazine, 1970, p. 56D)

Evidently, Minsky's prediction was overstated and did not come to pass. Soon after that, political positions related to AI research and development in the United States and Great Britain changed in the aftermath of British mathematician James Lighthill's report, which cast doubt on the ability of AI to move beyond simple problem-solving tasks and reach human-like reasoning capabilities. Consequently, funding for AI research from the British and U.S. governments was abruptly discontinued, and no momentous advances in AI could be noted during the next decade or so thereafter, a period of time in AI development known as the "AI Winter" (Haenelein & Kaplan, 2019; O'Neill, 2023). However, research in AI continued, and soon funding for AI projects resumed, culminating in the creation of expert system prototypes, the most prominent of which was

IBM's *Deep Blue* chess playing machine that defeated chess grandmaster and world champion Gary Kasparov in 1997. It did so via immense computational power at extremely high processing velocity, with observed system speeds in searches for positions and moves ranging from an average of 126 million to a maximum of 330 million positions per second, far outstripping Kasparov's ability to match such tremendous computational calculations (Campbell et al., 2002). Notably, this victory of a chess-playing machine over the most decorated chess master in the history of the sport came only a year after Kasparov defeated an earlier iteration of Deep Blue, attesting to the rapid advances in the development of ever more potent computational capabilities in humankind's drive to create intelligent machines. Despite the immense progress over time in enhancing their computational processing power, expert systems such as ELIZA, the General Problem Solver, and Deep Blue suffered from the absence of a critical quality specific to human, that of learning from, interpreting, and analyzing the data in order to adapt their responses in flexible ways. More specifically, expert systems relied on "collections of rules which assume that human intelligence can be formalized and reconstructed in a top-down approach as a series of 'if-then' statements" but "performed poorly in areas that do not lend themselves to such formalization" (Haenelein & Kaplan, 2019, p. 8). Instead, a shift was needed in how computers processed information that brought them closer to how the human brain functioned (Russell & Norvig, 2021), namely a closer replication of the pathways created through the synaptic connections among its hundred billion neurons in the process of learning, allowing humans to store, retrieve, and manipulate information at high speeds and in myriad circumstances.

Thus, the concept of artificial neural networks emerged in AI research that now holds the promise, or the specter, of AI becoming ever more powerful and, eventually, superseding the human brain's computational performance estimated at 10 quadrillion (10^{16}) calculations per second (Moore, 2014). Furthermore, it has been shown that "neural networks can be trained via evolution" (Adami, 2021), prompting claims that AI should be seen through a Darwinian lens that views AI systems as growing and learning through developmental processes of adaptive complexity driven by natural selection (Spector, 2006). The prospect of increasingly more complex, faster, and more efficient AI machines, able to surpass human intelligence, led computer scientist, futurist, and one of the pioneers of pattern-recognition technology, Ray Kurzweil, to proclaim that

"I set the date for the Singularity—representing a profound and disruptive transformation in human capability—as 2045. The nonbiological Intelligence created in that year will be one billion times more powerful than all human intelligence today" (Kurzweil, 2005, p. 122). Whether Kurzweil's prediction will materialize is yet another open question, but it is certain that, with the advent of artificial neural network, new advances in AI performance are on the horizon. Artificial neural networks allow AI systems or mechanisms to engage in what has been termed as machine learning or deep learning, allowing them to recognize complex data patterns, whether text, images, sound, or other types of data, to draw conclusions, generate insights and make predictions. It is through such advances that in 2015, *AlphaGo*, a program developed by Google, defeated the European champion at the board game Go. Echoing Deep Blue's defeat of Kasparov from only two decades prior but relying on thousands of times fewer calculations per second than Deep Blue, AlphaGo was trained to evaluate board positions more efficiently through "value networks" and select positions via "policy networks" in a combination of supervised learning from human expert games and reinforcement learning from self-play games (Silver et al., 2016).

These exponential developments in AI today form the basis for a multitude of applications, such as image recognition algorithms (e.g., Google Image Search, Facebook, etc.), speech recognition algorithms, virtual personal assistants (e.g., Amazon's *Alexa*, Apple's *Siri*, etc.), or autonomous vehicles (e.g., Waymo, BMW, Tesla, etc.). Most recently, generative AI and large language models, such as *ChatGPT* (Chat Generative Pre-Trained Transformer), have captured the attention of academics, policymakers, scientists, corporations, and general users alike, both for the opportunities they offer and the challenges they pose to human activity in the broadest sense. ChatGPT, created by OpenAI, uses artificial neural networks to analyze massive amounts of data from which it generates new content that replicates natural human language. Although earlier iterations of ChatGPT produced rather rudimentary descriptions of data, its most recent version, ChatGPT-4, can already outperform humans in certain areas (deWinter, 2023), opening the distinct prospect that future versions or similar models will evolve over time to higher levels of intelligence (Spector, 2006). Coming full circle then, while the fantasies of mythical narratives and sci-fi novels or movies perused above have yet to materialize in their full magnitude, given time, it is certainly possible to anticipate that some versions of autonomous, thinking, embodied AI that closely resemble humans may emerge after all. However, it remains an open question whether such forms of AI will eventually become endowed with self-awareness, consciousness, emotions, and intentionality to emulate humans and their apparent aspirations to create life in their image.

Typologies of AI

It can be surmised from the preceding historical overview that the evolution of AI has gone through different phases of development, not necessarily in an incremental fashion, but marked by periods of progress followed by stagnation at times. Nonetheless, this evolution has produced certain types of AI, which can be grouped into two broad categories, namely AI capabilities and AI functionalities. These categories can be divided further, with AI based on capabilities consisting of Narrow AI, General AI, and Super AI, while AI based on functionalities comprises Reactive Machine AI, Limited Memory AI, Theory of Mind AI, and Self-Aware AI. As AI systems have exerted an influence and will continue to increasingly infiltrate human activity in many domains, including education, it is worth reviewing them here briefly.

AI Based on Capabilities

Narrow AI. Also known as weak AI, narrow AI is currently the most common form of AI, with any other AI still existing in various stages of conceptual or theoretical formulation. It is considered narrow because this type of AI can be trained to perform a specific or single task at speeds that far exceed human capabilities, given that it relies on sheer computational power, but it cannot operate outside that narrowly defined task. In other words, "current AIs provide narrow solutions to specific problems; they aren't general problem solvers" (Loukides & Lorica, 2016, p. 3). Some examples noted above, such as ELIZA or Deep Blue, and even AlphaGo, with its use of artificial neural networks, represent narrow AI, as they are trained to self-learn and perform a specified task with high efficiency. AI expert systems have already made inroads into education, and their impact is only expected to grow as their adoption accelerates. Intelligent computer-assisted learning (ICAL), intelligent or adaptive tutoring systems, intelligent language learning programs, personalized learning systems, and similar expert systems have been integrated into curricula to enhance student learning (Sora & Sora, 2012). By harnessing vast amounts of data and combining formal and informal learning resources, expert systems may help provide students with personalized learning experiences. At the same time, expert systems can assist educators in creating adaptive learning materials to use in their teaching and in automating routine administrative and assessment tasks by relying on learning analytics and educational data mining processes.

General AI. General AI, or strong AI, is a largely theoretical form of AI, envisioned as a type of AI that mimics the workings of the human brain. However, even combining pattern-matching techniques, rule-based decisions, and exhaustive tree searches, all distinct forms of narrow AI, general AI does not amount to the functionality of human intelligence. Even the use of artificial neural networks, while computationally useful, cannot replicate the sophistication of human thought as much is still left unknown about how the human brain works. That is because "what humans appear to do better than any computer is to build models of their world and act on those models" (Loukides & Lorica, 2016, p. 4). As such, the promise or prospect for general AI is currently unclear, and its impact on education can only be discerned insofar as future advancements in the aspirational goals towards creating thinking, sentient, and emotional machines will translate from desideratum to palpable manifestations of it.

Super AI. Super AI is commonly referred to as artificial superintelligence or hyper-intelligence, like general AI, remains a strictly theoretical concept. If it were to materialize, this form of AI would possess the qualities of thought, reason, learning, judgement, and cognitive abilities that would far exceed any human capabilities. Further, super AI would likely be expected to be endowed with creativity and initiative, possibly evolving into a form of intelligence that can develop a value and belief system, and desires of its own. It is unclear how such a level of intelligence might be achieved, and it may linger in a utopian state for some time. It remains to be seen whether forms of superintelligence could someday work side-by-side or even replace humans in various domains of activity, including education.

AI Based on Functionalities

Reactive Machine AI. Reactive machine AI represents systems without the ability to store memories or previous experiences to inform decisions, rather, as their name suggests, they are designed to react to information or data they perceive. Since they are designed to handle a single task by processing and analyzing data present in the moment, they are unable to recall or recollect prior decisions, but merely react to a situation presented to them and predict the best course of action or next move from the data available to them at a given point in time. An example of a reactive machine is the chess-playing computer mentioned earlier, Deep Blue, which defeated the chess grandmaster, Gary Kasparov. Deep Blue used its massive computational power to calculate the positions of the chess pieces on the board in real time to predict its moves and those of its opponent without drawing from any previous memories or experiences. In the realm of education, reactive machines could come in the form of educational games, using decision trees or predetermined randomized algorithms to walk the learner through learning experiences that rely on user input.

Limited Memory AI. In contrast to Reactive Machine AI, this type of AI can recall past events and use some of its experiences to make decisions. Although it can use past- and present-moment data to decide a course of action, Limited Memory AI can only store a certain amount of data in its memory, but only for a short period of time. While it cannot retain past experiences in its data library over the long term, sustained training on increasingly larger data points may improve its performance. Self-driving cars or autonomous vehicles embody this type of AI, as the AI systems on which they operate constantly compare immediately available data from their environment with accumulated experiences over time to make limited decisions (e.g., navigational, directional, corrective, etc.) in steering the vehicle to its destination. In education, expert tutoring systems can be considered an instantiation of Limited Memory AI, as these guide the learner through responses based on user input and previously stored patterns of responses either from the same user or from norm- and criterion-referenced cohorts of learners. As another example, generative AI systems, such as ChatGPT, can also learn from previously stored information and experiences to generate responses that may help learners synthesize knowledge, although the risk of misuse cannot be understated. Learning analytics is another area in which Limited Memory AI is infiltrating educational processes, as the learning assessment data generated and then analyzed is stored in data systems that then allow for personalized predictions for learner performance improvement resulting from such analysis, leading to recommendations for individualized learning approaches that are adapted to each learner's unique needs and abilities.

Theory of Mind AI. Theory of Mind AI is an advanced AI class currently in the conceptual stage. This form of AI may be able to understand human emotions and thoughts and, therefore, affect how it interacts with those around it. This understanding of human traits would also allow this type of AI to understand that human emotions, sentiments, or thoughts are fluid or shifting depending on the environment in which they are contained. Consequently, Theory of Mind AI could simulate human-like interactions, but would not be able to generate emotions or thoughts of its own. Attempts at implementing Theory of Mind AI resulted in the 1990s with the development of Kismet by a team of engineers at the Massachusetts Institute of Technology. A robot head, Kismet, could recognize and mimic human emotions but could not follow gazes or convey attention to humans. More recently, Sophia, a similar robotic humanoid developed by Hanson Robotics, was also able to simulate human emotions, but as an improvement over Kismet, it could sustain eye contact, recognize individuals, and trace the movement of faces.

Self-Aware AI. Like Theory of Mind AI, Self-Aware AI is a hypothetical or purely theoretical class of AI systems that would exhibit superintelligence or Super AI (see above). Should it ever materialize, this kind of AI would also understand human emotions and thoughts and be able to interact with humans as virtually indistinguishable from the latter. Such systems, applications, or devices would not only be able to replicate or simulate human traits and behaviors but would also obviously surpass any biological human intelligence. Furthermore, as the name implies, they would have the capacity to understand and contemplate the meaning of their personal existence as sentient beings, and would possess emotions, form beliefs, and express their own needs. In essence, self-aware AI would likely evolve into claiming personhood status and demanding roles, rights, and responsibilities in society currently attributed to human beings. Should they become a reality, both Theory of Mind and, especially, Self-Aware AI would have significant implications for society in general and education. It is possible to envision a future in which such AI would teach and learn side-by-side with humans, but it is far from clear what the relationship between AI sentient and human beings in educational settings would look like. Such AI systems could either collaborate and work in partnership with humans, perhaps assisting humans in their teaching and learning activities, or they could rival and even replace humans as they create parallel societal and educational structures that play to their advantage or serve their needs. Because, in principle, they would replicate and develop on their own, independent of the computer algorithms that humans used to create them, these forms of AI could reproduce without human intervention, leading either to harmonious or conflictual relationships with humans and generating tension in society. Such scenarios certainly carry risks and consequences that are difficult to anticipate and pose critical questions of AI ethics both at the societal level, in general, and in education, in particular.

A HISTORICAL OVERVIEW OF AI IN EDUCATION

It can be argued that the history of AI as a modern technological creation is virtually coterminous with its evolution within the field of education. The precise point in time at which the emergence of AI intersected the field of education is disputed, but estimates for a timeline range from as early as the 1960s or 1970s (Guan et al., 2020) to slightly over three decades ago (Zawacki-Richter et al., 2019). In fact, an entire field dedicated to AI in education has emerged over the past nearly three decades to examine and explore the ramifications of AI for education. During this time, AIED, as the field is known, has matured (Holmes, 2023) and is now professionalized to the extent that its scientific community can claim membership in the *International Artificial in Education Society* (IAIED), which identifies as "an interdisciplinary community at the frontiers of the fields of computer science, education, and psychology" (International IAED Society, n.d., para 1). Since its inception on January 1, 1997, the Society has grown to over 1000 members from 40 countries, has been bringing together its research community to its annual conferences, and has been advancing research in the field through its scholarly publication, the *International Journal of Artificial Intelligence in Education* (IJAIED).

Consequently, AI and education have been inextricably linked, as AI researchers have attempted to elucidate how learning abilities exhibited by humans could be adapted to AI systems or machines, and conversely, they have sought to develop machines that could understand human thought and information-processing capabilities. In many ways, learning theories in education, particularly guided by cognitivist approaches, have influenced the development of AI, as human designers have aimed to create AI systems that could replicate human cognitive abilities. Moreover, many contemporary AI breakthroughs have occurred in research and development projects spearheaded by academics in higher education institutions. While, at the dawn of modern scientific research in AI, the prevailing paradigm informing educational provision and curricular development rested in behaviorism, AI scientists were instead focused on the information-processing traits of the human brain, in order to illuminate the avenues by which they could develop AI that emulated the cognitive processes humans possessed. As Doroudi (2022) suggests,

> Early artificial intelligence pioneers were cognitive scientists who were united in the broad goal of understanding thinking and learning in machines and humans, and as such were also invested in research on education. The point is not just that they were cognitive scientists whose work had implications for education, but rather that these researchers were also at times directly involved in education research and had a significant impact on the course of education research. (p. 4)

Hence, much of the initial impetus for conceptualizing the theoretical underpinnings for the development of AI stemmed from the foundation of *information-processing psychology* spearheaded by the aforementioned Dartmouth Workshop proponents and organizers, Herbert Simon and Allen Newell, both of whom were professors at Carnegie Mellon University, along with J.C. Shaw. This branch of psychology defined the mechanisms by which humans engage in problem-solving and postulated that, as a consequence, learning amounted to a process "of gradually accumulating the various production rules necessary to solve a problem" (Doroudi, 2022, p. 6). Primarily due to their work in both the cognitive sciences and AI, information-processing psychology provided an alternative to the behaviorist paradigm of the 1950s in education, and by the 1970s, it became the predominant learning theory informing educational processes. Therefore, they, their later associates, and ensuing cohorts of cognitive scientists interested in AI attempted to infuse AI machines with the principles, algorithms, rules, and mechanisms underlying problem-solving. These early efforts translated into intelligent tutoring systems such as the Logic Theorist and Merlin (Moore & Newell, 1974), the latter intended to teach graduate artificial intelligence.

However, in the 1980s, critics of the information-processing approach, or cognitivism, proposed a countervailing theory that saw learning as a context-dependent, situated process, not one limited exclusively to the rules of problem-solving postulated by Simon and Newell. Such critics argued that cognitivism, by itself, could not explain the situatedness of the learning process (Resnick, 1987), a perspective influenced by sociocultural theories drawing on Lev Vygotsky's work that viewed learning and construction of knowledge as a consequence of the learners' interaction with their cultural and social environment. It is also apparent that situativism arose in reaction to and was influenced, at least in part, by work its proponents had conducted in AI development, who had identified the limitations of traditional AI work until that time. For instance, one of the core proponents of the situated perspective in education, John Seely Brown, developed some of the earliest intelligent tutoring systems (Brown et al., 1989). It is, therefore, no coincidence that these innovations in AI systems eventually led to situativism gradually superseding, though not supplanting, cognitivism in education, becoming the prevailing perspective in learning sciences and, consequently, in the educational domain.

Notwithstanding which learning paradigm prevails in education, there is rising concern about the conceptualization of AI as a purely technical phenomenon, the implementation of which in education is context independent, benign, and intrinsically beneficial for the teaching and learning processes it intends to advance. While the debate over which learning theories would or should inform the continued progress in the evolution of AI technologies, such as adaptive or intelligent tutoring systems, interactive learning environments, learning analytics, natural language processing applications, and generative or large language models, the enthusiastic embrace of these technologies (Roll & Wylie, 2016) may have to be tempered by important ethical questions that arise from scenarios in which their actual utility and benefit remain undisputed (Holmes et al., 2022b). Consequently, the problematique of AI as an objective arbiter or enhancer of learning attainment is currently challenged as myopic, given that competing interests, be they economic, commercial, or political, may impede their meaningful and equitable utilization for the benefit of teachers and learners. As noted earlier in this chapter, AI is poorly defined and far from being a concrete or distinct technological system or apparatus. Rather, AI consists of incoherent and disparate sets of technologies at various stages of evolution and exhibiting different capabilities, typically in narrowly specified domains. Moreover, its development results from the accrual of knowledge about machines replicating human cognition, computational theory advancements and hardware development embedded in social and historical contexts surrounding humanity's needs and goals. As such, different stakeholders understand AI in education differently, with unresolved and complex implications for the educational sector. Accordingly, depending on from which vantage point vested actors view the role of AI in education, they may conceive it "as a methodology for academic research to better understand learning and achieve practical impact on learning and education outcomes; as a potential source of profit for industry; and as political rhetoric used to demand educational reforms" (Williamson, 2023, p. 3). Any or all of these three perspectives combined can have both favorable and deleterious effects on the beneficiaries they intend to support, namely the learners themselves.

Thus, in recent years, international organizations have rushed to propose and formulate principles for the ethical uses of AI, especially in education, to protect learners' human rights and dignity, whether in formal or informal settings. For instance, in its *Recommendation on the Ethics of Artificial Intelligence*, the United Nations Educational, Scientific and Cultural Organization (UNESCO) cautions about the profound impact AI may have in challenging human agency and practices, stating that

> AI systems raise new types of ethical issues that include, but are not limited to, their impact on decision-making, employment and labour, social interaction, health care, education, media, access to information, digital divide, personal data and consumer protection, environment, democracy, rule of law, security and policing, dual use, and human rights and fundamental freedoms, including freedom of expression, privacy and nondiscrimination. (UNESCO, 2021a, p. 10)

In this context, UNESCO is expressly concerned with the impact of AI systems on education "because living in digitalizing societies requires new educational practices, ethical reflection, critical thinking, responsible design practices and new skills, given the implications for the labour market, employability and civic participation" (UNESCO, 2021a, p. 10). It invites its member states to collaborate with international organizations, educational institutions, and other public or private actors to ensure that their citizens receive adequate AI literacy skills in order "to empower people and reduce the digital divides and digital access inequalities resulting from the wide adoption of AI systems ... especially in countries and in regions or areas within countries where there are notable gaps in the education of these skills" (UNESCO, 2021a, p. 33). Further recommendations in the area of education relate, among others, to: protecting human rights in the provision of educational programming impacted by the advent of AI; creating capacity for the implementation of AI technologies in learning and teaching; safeguarding against the increased exploitation or extraction of sensitive individual data via AI systems in educational processes; ensuring the protection of vulnerable people and their inclusion in educational programming supported by AI technologies, and; preserving and promoting multilingualism through the use of AI to enhance the provision of education in local, indigenous, and/or marginalized languages.

Concerns echoing those raised by UNESCO relate to the expansion of AI in education, a phenomenon referred to as platform colonialism (Sujon, 2019). This is a process by which large commercial tech companies, such as Google, seek to promote, monopolize, and profit from adopting their products in education. Driven by the mantra of efficiency, convenience and the illusion of creating individualized learning experiences, the digital learning platforms implemented in classrooms around the world provide instead standardized, monolingual, uninspiring, and decontextualized learning scenarios that, rather than catering to the unique needs of its users in various locales, exacerbate existing cleavages between the Global North and the Global South by favoring predominant Western-centric ways of learning over local approaches to learning. With the advent of AI, powerful companies in this industry, aided by their global reach, may seek to capitalize on AI systems' computational and cognitive power to impose AI tools produced in the Global North on educational systems, classrooms, teachers, and learners in the Global South. In fact, in a report commissioned by the Council of Europe on the impact of AI in education viewed from the perspective of human rights and democratic values, Holmes et al. (2022a) interrogate such monopolization of knowledge on the part of global tech multinationals, proffering that

> AIED colonialism might constitute the physical adoption by the decision makers of AIED tools created in one context in other places. Territorial gains can be made across schools as institutions, or as segments of entire state education systems, where a country or region adopts a single product across all its schools. (p. 45)

Not only is this approach unethical or unconducive to fostering local knowledges, but it also creates a dependency on AI technologies produced for users or audiences unrelated to the populations of the Global South. Instead of supporting the cultural and educational diversity of the non- or under-industrialized world, the uncritical adoption of AI systems in education, as previous waves of technology adoption in education have shown, may only reproduce the biases already inherent in data-driven learning assessment mechanisms, educational platforms, learning analytics, or intelligent tutoring systems built on learning performance and cognitive assumptions rooted in Western learning theories. In the coming years, a critical stance toward the indiscriminate and unexamined adoption of AI tools in education will remain an important function of researchers, practitioners, and policymakers in creating the regulatory frameworks on AI in education. Their collaboration both within and across the Global North and the Global South will be paramount in safeguarding the humanity and dignity of educators and learners from the worst effects of the commercialization of AI in education.

AI Implications for CIE

As a cognate field and sub-domain of education, comparative and international education (CIE) is exposed to the same, if not a greater, extent to the visions, contemplations, and potentialities of AI developments sweeping through education at this point in time. Given its inter/pluri/multidisciplinary nature, CIE is influenced in myriad ways by the conceptualizations and empirical representations of the instrumentality of AI in education. As argued by Salajan and jules (2021), CIE may be conceived of as a complex educational assemblage, which they define as "polymorphic and multiscalar arrangements of educational polities, systems, and mechanisms, bound together by symbiotic and synergistic relationships, driven by shared purposes, mutual interests, and common responsibilities" (p. 150). They further highlight the vulnerability and permeability of the field to the influence of digital technologies, "particularly as Big Data flows create and consolidate new informal and formal pathways of bureaucratic and socio-human interconnectivity between elements of educational polities" (Salajan & jules, 2021, p. 150).

Given its varied geometry, the global ethical considerations on AI in education discussed above, and its focus on the comparative examination of educational systems, processes, and practices both within and among countries, regions, or other (non)territorial arrangements, CIE is uniquely positioned within the larger scholarly, practice-oriented, and policy domains to both elucidate and critique the role of AI in education. Thus, CIE can contribute to several areas of interest for comparativists and scholars in cognate fields interested in educational comparison. For example, current conversations on the advent of AI in education may examine its role in advancing scholarship in the field of CIE, in an era in which the field is reckoning with its colonial past. (Re)constructing new understandings of the field's colonial legacies and decolonizing the act of comparison to embrace new intellectual commitments from Indigenous and BIPOC traditions also entail a commitment to resist succumbing to AIED colonialism. These two examinations need to work in tandem, if comparison can make a moral claim to inclusivity, openness, and equity for onto-epistemological assumptions enriched and transformed by cosmologies from the Global South. Relatedly, CIE needs to interrogate the (false?) promise of AI in promoting a more just and equitable use of (new) research methodologies and ask whether AI can level the playing field for scholars, teachers, and practitioners so it is inclusive of heretofore marginalized onto-epistemologies and cosmologies or whether it will reinforce the domination of western scientific thought in CIE. While AI may hold the potential to infuse the field of CIE with new tools to support research, teaching, practice, and policy directions that reaffirm the role of comparative perspective taking in the examination of interconnected educational phenomena, such assumptions should not go unquestioned, as AI-enhanced research tools may also entrench biases in the act of comparison, rather than support a holistic view of comparative approaches that blend the established with emerging methodologies that inform practice adapted to localized needs. As the current debate at the nexus of research, practice, and policy with and on AI in CIE is in its infancy, it remains to be seen how the evolving nature of AI tools will inform the act of comparison. How CIE might (re)envision its role as an arbiter for ethical, socially just, inclusive, and holistic research practices with and around AI is the object of the conversation in the remaining chapters of this book.

CHAPTER 4

Beyond the Anthropocene: Ethics, Equity, and Responsible Use of AI in CIE

Abstract This chapter explores the ethical dimensions of integrating AI into the field of CIE. By situating the emergence of AI as a significant new player in the educational landscape within the context of the Anthropocene, the chapter sheds light on the ramifications of this new actor for environmental sustainability, human agency and governance, and equity and access. The chapter also examines the risks associated with digital coloniality within CIE, scrutinizes the ethical considerations linked to the use of AI in the field, and illustrates some of the advantages that AI advancement offers to researchers and practitioners. While acknowledging the potential benefits of AI for CIE, the chapter underscores the need for ethical guidelines to ensure alignment with principles of justice, equity, and the field's decolonizing efforts. It advocates for a critical assessment of AI narratives and applications and stresses the importance of shared responsibility between users and creators in addressing ethical concerns. Ultimately, the chapter emphasizes the significance of collaborative research and practice in navigating the ethical challenges posed by AI, with the aim of fostering a future where education is both technologically advanced and ethically grounded.

Keywords Ethics • Equity • Risks • Benefits • Bias • Anthropocene Era • Digital coloniality • Sustainability

S. M. S. Curtis et al., *The Technological-Industrial Complex and Education*, https://doi.org/10.1007/978-3-031-60469-0_4

Introduction

The question that guides this book, *how can CIE scholars and practitioners use AI to decolonize research and practice in CIE?*, is an ethical question. In a recent paper on generative AI and the future of education, UNESCO's (2023a) core message is that we need to ensure AI tools are appropriately integrated into educational learning settings on our terms. However, a study conducted by UNESCO (2023b) showed that only 10 percent of 450 schools and universities worldwide had developed institutional policies and guidelines regarding the use of generative AI. Undoubtedly, AI can open new educational horizons, but we need to work to ensure that the risks do not outweigh the benefits.

This applies to CIE as well. The field should establish ethical guidelines for using AI to ensure its application leads to more justice and equality rather than the opposite. It is crucial to recognize that technology, including AI, is a product of human activity shaped by historical context. It is not a neutral or objective tool brought by "progress"; instead, it is intricately connected with systems of meaning, value, ethics, and power that influence its development and direct its use (Creutzig et al., 2022; Werthner et al., 2022). Therefore, implementing AI should not be automatically considered positive or beneficial. As Means (2018) highlights, "technology is always embedded within ideological assumptions and can have both dystopian and utopian potentials. Therefore, technology narratives and applications can never be taken at face value and require critical assessment" (p. 98). Moreover, considering the colonial roots of CIE and its current involvement in a decolonial project, as discussed in the preceding chapters of this book, the field must carefully assess the risks and benefits of integrating AI into the study of education across countries to avoid perpetuating new forms of colonial influence.

CIE entails the study of the world's educational systems from a comparative perspective to discern what can be learned from approaches in various educational contexts and how these approaches might be applicable to other educational systems. The advent of advanced technologies, particularly AI, has been a game changer for research, teaching, and learning methods. In its simplest terms, AI is the ability of machines or computers to execute tasks that would typically require human intelligence, such as learning, problem-solving, and decision-making. It is within this context and at the intersection of deglobalization (and the retreat towards regionalization) and the rise of the Fourth Industrial Revolution that this chapter seeks to offer cautions and underscore some benefits of the use of AI in the field of CIE.

As of 2021, more than 30 countries have released national AI policy strategies that articulate plans and expectations regarding how AI will impact different policy sectors, including education (Schiff, 2022). UNESCO Miao et al. (2021) note that AI has the potential to tackle significant challenges in education, revolutionize teaching and learning methods, and facilitate advancements towards SDG 4. However, they have also warned that the fast technological progress introduces numerous risks and challenges that have surpassed current regulatory frameworks and policy discussions. Intending to explore the risks and benefits that AI poses within the field of CIE, this chapter considers the following inquiries: How can decolonization coexist effectively with platformization/connectivity, emphasizing a pluriverse (Escobar, 2020) rather than fostering homogeneity and a singular-world model? How can CIE scholars address and include forms of knowledge that resist digitalization? And, crucially, how can the field ensure that the integration of AI does not inadvertently introduce new forms of domination, advantage, or colonial influence?

To address these questions, this chapter begins with a review of AI during the Anthropocene era. The chapter then delves into the risks of digital coloniality for CIE, given its ongoing goal to decolonize the field. Next, it scrutinizes the ethical considerations associated with AI's utilization in the field of CIE. It begins by drawing distinctions between the application of AI for researchers versus practitioners in the field, shedding light on the cautions for both groups when deploying AI for studying foreign regions. The subsequent section showcases some of the advantages or benefits that the advancement of AI has yielded for both researchers and practitioners in the field. Finally, the chapter wraps up with conclusions and outlines future directions for the ethical implementation of AI in CIE.

AI ETHICS IN THE ANTHROPOCENE ERA

The term "Anthropocene" is used to describe both a specific period of time within the history of humanity and a critical moment that has created the opportunity to rethink the human-nature relationship and de-center humankind (Lövbrand et al., 2015). Despite the myriad ways of defining the Anthropocene, a shared consensus among perspectives is that this period was initiated when humans started influencing the climate in manners that jeopardized our survival on Earth. In the Anthropocene era, humans are the main force shaping the planet, which involves dehumanizing individuals in the name of progress and treating the nonhuman realm as if it were human property (Sutoris, 2022). Because climate change was the first sign of the Anthropocene, and because of the severe implications environmental crisis can engender (literally our extinction), initial responses emerged from scientists to stop or deaccelerate the process that can lead to our annihilation. As Crutzen (2002) indicated, in the era of the Anthropocene, scientists and engineers confront the challenging responsibility of leading society towards sustainable environmental management. However, permanent and real solutions to the crisis of the Anthropocene cannot only originate from reorienting our practices towards more green ones. All fields need to rethink "long-held assumptions about the autonomous, self-sufficient human subject that begins and ends with itself" (Lövbrand et al., 2015, p. 213). Within this new era that calls for a post-human perspective, AI, a new non-human actor, has come to challenge and add a new layer of complexity to the relationship between beings and non-beings, bringing both hope and concern. Creutzig et al. (2022) offer an analysis of the effects of AI in the Anthropocene crisis, focusing on AI's impact on three elements: "environmental protection and planetary stability, human agency and its governance, and equity and access" (p. 481). These three effects expose key ethical concerns about the role of AI in society.

Regarding the effects on sustainability, Creutzig et al. (2022) clustered the environmental impacts of digital technologies into two groups: direct effects, which stem from the digital hardware and infrastructure's life cycle, and indirect effects, that arise from the consequences of specific technology usage, including behavioral and systemic outcomes. The direct impacts include greenhouse gas emissions, consumption of water, the use of resources to produce technological devices (e.g., the extraction of raw materials like copper, lead, tin, antimony, indium, gallium, germanium, and ruthenium), and the high-level electronic waste that comes from designing products with short life spans (Creutzig et al., 2022). The indirect environmental consequences of digital technologies result from the continuous data that is generated, which fuels the exponential expansion of specialized device production and increased consumption. This heightened demand is mainly perpetuated by the widespread integration of these technologies into every facet of life and the easiness of the purchase process. Furthermore, digital technologies foster an anthropocentric and individual mindset that increasingly prevents individuals from acknowledging and comprehending the severe environmental implications of technological development (Bowers, 2014). As outlined in Chap. 2, the environmental consequences of AI development are rooted in the reproduction of colonial logics, based on oppression, exploitation, and dispossession.

Means (2018) exposes the issue of "solutionism" brought by technological advancement—the belief that all societal issues, including climate change and social inequality, are technical issues that can be solved through data platforms and technology. The main problem of solutionism is that it constrains social imagination. According to Means (2018), "when society is reduced to economic and technical calculations, our collective ability to define how complex problems are posed, understood, debated, and addressed is eroded" (p. 2). Despite the challenges digital technologies pose to the environment, many researchers believe that "digitalization also holds a substantial but unrealized potential for stabilizing the planetary trajectory" (Creutzig et al., 2022, p. 490). AI is crucial in unlocking this potential, as it can help address sustainability challenges and develop informed strategies for responsible global stewardship. Digital technologies, data science, computer science, and machine learning are pivotal in climate and Earth sciences. These technologies facilitate the vast and highly detailed collection of raw data through sensors, satellites, and similar means, advancing scientific discovery. For example, Creutzig et al. (2022) illustrate how "remote sensing approaches combining machine learning techniques and satellite images have been used to map and help plan the extension of solar photovoltaic installations at a global scale" (p. 490). Another example is the use of AI to identify and regulate energy consumption patterns and carbon emissions to inform policy measures (Tamburrini, 2022). Thus, the dual nature of AI is evident, presenting both negative impacts and substantial potential benefits in addressing environmental challenges. However, this is just one aspect of the broader range of risks and benefits associated with AI. Urgent attention is needed for a comprehensive examination of potential drawbacks and advantages, coupled with developing a global strategy for the responsible use of AI to benefit society and mitigate harm.

Another ethical concern of AI raised by Creutzig et al. (2022) pertains to human agency and governance, arising from the increasing generation, processing, and storage of data in the current digital era. This data is concentrated within internet firms, mainly social media platforms and digital services. While these platforms provide social benefits like enabling political and social organizations and supporting local networks, they also raise three main concerns. First, data accumulation can lead to unchecked monopolies, potentially limiting innovation and perpetuating power imbalances. Second, these powers can easily predict and manipulate people's behaviors, choices, and beliefs for their own profit. Finally, the capacity to manipulate user behaviors can threaten democracy (Creutzig et al., 2022). Chapter 6 of this book provides an in-depth exploration of the intersectionality of AI, governance, and the field of CIE.

The third and final concern Creutzig et al. (2022) address is equity and access. The reality is that "digitalization is developed unevenly both within and across countries globally" (Creutzig et al., 2022, p. 485). Given the unequal distribution of resources and wealth across countries, equity is a primary concern when considering implementing AI across fields, including education. Another significant concern in terms of equity is the impact digitalization has on how value is generated and distributed within and among economies. For instance, it has led to a shift in the balance between the value created by services versus physical products. Additionally, it alters trade patterns by reducing transaction costs in logistics. This diffusion of digital technologies gives rise to issues related to distribution, particularly concerning labor demand, wage disparities, the digital divide (unequal access to technology), and the disproportionate environmental impact on low- and middle-income countries (Creutzig et al., 2022).

As Creutzig et al. (2022) state, the question is: Is digitalization beneficial or detrimental to developing nations? On the one hand, it offers the potential for these countries to catch up and compete effectively in the global digital economy. For instance, in the early 2000s, the internet economy raised hopes that outsourcing professional business services could create job opportunities in countries like India, which possess favorable conditions, such as a large, English-speaking workforce. On the other hand, digitalization leads to labor-saving automation in factories and more efficient business processes. This may result in challenges for low- and middle-income countries seeking employment opportunities for their growing populations. Simultaneously, the decreasing significance of labor compared to capital weakens these countries' competitive advantages based on low-cost and abundant labor (Creutzig et al., 2022). Furthermore, the benefits of digitalization are currently concentrated in developed countries. A substantial portion of the value created in ICT manufacturing and the installation of global robots is centered in a few nations. Similarly, the world's largest online platforms are mostly headquartered in the developed regions of the Global North, leaving those in South America, Africa, and other Global South disadvantaged regions. Therefore, while digitalization has the potential to benefit countries in the Global South, achieving this will likely require breaking free from global structures of dependency (Creutzig et al., 2022).

In essence, it is imperative to investigate whether, amid the environmental challenges, the exploitative nature of capitalism and neoliberalism, and the enduring impact of colonialism and imperialism, AI amplifies these issues or potentially signifies a move towards a more equitable and just society. Examining the ethical dimensions of employing AI in CIE can provide perspectives on this question and inspire both theoretical discussions and practical measures within the field. The remaining of the chapter delves into these ethical considerations. Additionally, we propose that reflecting on the colonial dynamics that underpin the bases of AI logics requires immediate attention in the field of CIE (Muldoon & Wu, 2023). How does a field committed to decolonizing education systems navigate the use of AI tools—considering the pervasive presence of digital coloniality within the AI field?

Lessons for CIE: Digital Coloniality and its Three Algorithmic Harms

Before we can speak of ethics and consider the ethical use of AI in education, we must broaden the discussion on coloniality and decoloniality that we began in Chap. 2. Chapter 2 has provided the foundation for comprehending the colonial principles that underlie the AI field and the effects these principles exert on the field of CIE. The latter is entrenched in colonial legacies but has undertaken a project of self-decolonization in more recent decades. This section discusses the challenges that engaging with a field characterized by coloniality (the field of AI) poses to CIE—considering its goal to "delink" (Mignolo, 2007) itself from the Western logics of colonization. As described in Chap. 3, the convergence of AI and the field of education occurred around three decades ago. The emergence of a distinct field, known as AIED, took place in 1997. Since then, a connection between AI and education has been established. However, AI did not originate within the educational context, and the field of education has no control over the development of AI. It can modify its tenets and use the tools that are beneficial for education research, teaching, and learning, but it cannot change AI's positivist and colonial logics.

The exclusion of many individuals, especially from the Global South, from discussions about the future of the digital society is a deeply rooted systemic problem. This disparity is often called "digital coloniality" (Bon et al., 2022) and has implications for the ethical use of and access to AI. As Bon et al. (2022) explain, "if we consider the digital society to be an image of the physical world, it will have inherited, along with other aspects, historical patterns of inequality. These patterns are referred to as 'coloniality'" (p. 62). Subsequently, from a decolonizing perspective, the primary concern revolves around determining who is defining technology and who has the decision-making power in digital development, emphasizing the necessity for collaborative efforts on AI development—especially in the most marginalized countries from the Global South. Collaboration and co-creation in building AI is the only way to promote a more participatory and democratic digital society that enables the emergence of a pluriverse, rather than perpetuating colonial logics (Bon et al., 2022). The pluriverse is defined by Escobar (2020) as a world where many worlds coexist, involving "an unambiguous refusal of the ontological imperative to be in a particular way" (p. xxii).

Another challenge in the decolonization of the field of AI involves the platformization or digitization of educational resources. The scarcity of resources for digitization and unequal access to technologies both present obstacles to the development of digital educational tools that avoid perpetuating colonial influences. Furthermore, developers need to ask if it is beneficial to put all knowledge into digital form, especially when it comes to indigenous knowledge that has already been objectified and misused in research, as Smith (1999) points out. AI developers should also ascertain whether the various communities worldwide endorse the digitalization of their resources and knowledge. The adverse outcomes observed in various AI applications globally, such as facial recognition, predictive policing, resource allocation, changes in labor practices, and healthcare diagnostics, are not coincidental. They stem from persistent and systematic mistreatment that can be traced back to the colonial project (Isaac et al., 2021). The field of AI is not only not exempted from the enduring colonial legacies, but it also actively participates in its reproduction. According to Isaac et al. (2021), it is especially urgent to confront three types of algorithmic harm present in AI: algorithmic oppression, algorithmic exploitation, and algorithmic dispossession.

Algorithmic oppression refers to the privilege some social groups have at the expense of other unprivileged or vulnerable groups. These are maintained through automated, data-driven, predictive systems, such as predictive policing and facial recognition. These tend to be based on unrepresentative datasets and reproduce social injustices. Algorithmic oppression is often called algorithmic bias (Isaac et al., 2021). While algorithmic oppression becomes evident in AI's implementation or production phase, algorithmic exploitation and dispossession arise during the research and design stage. Exploitation is most noticeable in the context of labor rights. The substantial datasets needed for AI system training are annotated by human experts, commonly referred to as "ghost workers." These tasks are usually performed in regions with limited labor laws and workers' rights. As Isaac et al. (2021) expose, "algorithmic exploitation construes people as automated machines, obscuring their rights, protections, and access to recourse—erasing the respect due to all people" (para. 4). Finally, algorithmic dispossession consists of concentrating the power, assets, and rights within a minority group. In the realm of algorithms, this might materialize in technologies that restrict or hinder specific modes of expression, communication, and identity (e.g., content moderation labeling queer slang as harmful) or through institutions influencing regulatory policies. It is essential to question who is being protected by these regulations and who defines and enforces them (Isaac et al., 2021). Algorithmic oppression/bias, exploitation, and dispossession need to be addressed if we want to develop AI tools that divert from colonial logics and practices and support a pluriversal world instead. The first step towards rebuilding the AI field is thinking about what we want to achieve in the future and how we measure success in achieving this future. In addition, AI developers, researchers, and practitioners should increase their understanding of the societal consequences of their work (Isaac et al., 2021).

An example of the effort from researchers and experts in AI to move towards decolonialized AI field is the *Decolonizing AI movement* that crafted the "AI Decolonial Manyfesto."[1] The group of around two dozen scholars developed over 16 months during the COVID-19 pandemic a Manifesto representative of a "shared desire to move beyond Western-centric biases in isolation driving global technology as we move forward into the algorithmic era" (para. 1). Countless other scholars have also been actively criticizing and reflecting on digital colonialism, contemplating ways to revise and reconstruct the field of AI to prevent it from

---

[1] See the Manifesto website here: https://manyfesto.ai/index.html. If you are interested in signing the manifesto, click on this link https://forms.gle/nrgwxnCWpULRAXVr8.

strengthening even more the "colonial matrix of power" (Quijano, 2007) that characterizes our current world-system—that has been shaped by imperialism and colonialism. Muldoon and Wu (2023) call for a "critique [of] the regimes of global labour exploitation and knowledge extraction that are rendered invisible through discourses of the purported universality and objectivity of AI" (p. 79). They present three arguments that emerge from this analysis and critique: that the world continues to function through global relationships that involve labor exploitation and knowledge extraction, that the development of AI has strengthened the international digital labor divide increasing the inequality gap, and that AI also reinforces the hegemonic knowledge production (Muldoon & Wu, 2023).

How Can CIE Respond to Digital Coloniality?

When it comes to the implementation of AI in the field of CIE, researchers and practitioners need to be aware of the colonial logics inherent in AI. This awareness is essential to exercising caution and selectivity. If a critical objective is to decolonize the field, the incorporation of AI should not be approached mechanically or unconsciously. When considering some of the goals of CIE as a field dedicated to the decolonization of education, it is imperative for CIE to, at the very least, acknowledge and respond to three conflicting characteristics of AI.

First, AI, primarily developed in the Western world by predominantly white men, has the potential to marginalize alternative modes of knowledge, perpetuating a lack of diversity in the field and excluding other cultural perspectives. Scholars like Birhane (2019) and Murphy and Largacha-Martínez (2022) argue that this has resulted in a phenomenon denominated "tech evangelism," contributing to the dismissal of other customs and values—a manifestation of what they describe as "algorithmic colonization." This raises critical questions for CIE as it grapples with the implications of AI in its pursuit of decolonization. A noteworthy example surfaced during the American Educational Research Association (AERA) in 2023, where the limitations of AI in education were discussed after some scholars presented different AI educational tools that were being used in diverse settings. In response to questions from the audience, most discussants agreed upon the importance of understanding the origins (including the developers and data used) of these technologies to understand and address their evident limitations. These are valuable discussions that need to happen regularly in CIE as AI continues to develop and grow in education.

Second, in the realm of AI, a perception of autonomy prevails, with algorithms seemingly capable of functioning independently without human involvement. This autonomy, however, erases local considerations, as algorithms are positioned as all-encompassing problem-solvers. The concern arises from the elevated status of these algorithms, which are considered almost like "gods," and their origins and underlying logics are rarely challenged. As Murphy and Largacha-Martínez (2022) note, this lack of scrutiny leads to widespread unawareness of the ongoing manipulation and the diminishing agency experienced by individuals because of AI. This raises important questions when incorporating AI into CIE research and practice. How does CIE grapple with individuals' unique contexts and potential alienation in the face of AI utilization? Additionally, how does CIE navigate the challenge of countering an overconfidence in the presumed "objectivity" and "accuracy" of algorithms? These inquiries underscore the critical examination required for integrating AI in CIE research and practice and its implications for individual agency and contextual understanding. In this context, the concept of autonomy inherent in AI development is very problematic because it can be traced back to the colonization process, through which colonizers appropriated the autonomy of the colonized. In other words, history has shown that nothing beneficial can emerge from conferring, or having stolen, all our decision-making autonomy to another entity (human or non-human). In Mhlambi and Tribelli's (2023) words, "consistent with its problematic conception and historical application, the principle of autonomy within Western derived AI ethics is a mismatch between the needs of those likely to be marginalized by AI" (p. 868). This flawed understanding of how autonomy in AI should work could be offset if developers of AI would consider more diverse philosophies and belief systems. For instance, Ubuntu philosophy considers the universe as organically and intrinsically interconnected, meaning that the individual cannot be without the community or the community without the individual. This ontological co-existence undermines the idea of building autonomous systems, as autonomy is considered weak and limited (Mhlambi & Tribelli, 2023).

Finally, a third noteworthy characteristic of AI that requires careful consideration from the field of CIE is its inherent positivism. The field of AI is deeply rooted in a scientific and idealized knowledge perspective, as Mohamed (2018) and Murphy and Largacha-Martínez (2022) note. This positivistic orientation perceives reality as holding an objective truth waiting to be discovered. CIE must think of effective strategies for countering such a positivistic view. Embracing participatory approaches, amplifying the voices of diverse communities, and valuing pluriversality are key avenues the field needs to explore and amplify. However, mere participatory practices and design might not necessarily dismantle colonization. Simply soliciting input, a common strategy used in the AI field to enhance "participation," falls short of decolonizing the field. What is imperative, and where CIE can contribute significantly to its own decolonization efforts while employing AI, is the adoption of community-based approaches. As Murphy and Largacha-Martínez (2022) assert when talking about an approach to decolonizing AI, "local knowledge and control are the centerpiece of this approach" (p. 11). Despite the significance of considering how the colonial legacies of AI intersect with the decolonizing goals of the field of CIE, the ethical considerations go beyond the complexities of this relationship.

As we proceed through the upcoming sections of this chapter, our exploration encompasses the cautions that CIE must follow for the ethical use of AI in research and practice and the notable advantages the field has gained from developing and implementing these tools in recent years. This examination is situated within the framework of social justice, delving into how the use of AI can either perpetuate existing disparities or act as a catalyst for more equitable educational outcomes.

USING AI IN CIE: INEQUALITY, BIASES, AND OTHER CAUTIONS

The discourse on ethics within digital development and, more recently, the rapid advances in AI are neither new nor scarce. Countless scholars and practitioners across diverse fields have expressed concerns about technological advancements that are becoming dehumanized and posing numerous harms for individuals (Bowers, 2014; Close et al., 2023; Crawford, 2021; Creutzig et al., 2022; Inverardi, 2022; Isaac et al., 2021; Lee, 2022; Richardson et al., 2023; Russell, 2022; Werthner et al., 2022). These scholars have noted that despite its contribution to solving significant problems, there are numerous shortcomings associated with the uncontrolled development of AI, including, for instance, "echo chambers and fake news, the questioned role of humans in AI and decision-making, the increasingly pressing privacy concerns, and the future of work" (Werthner et al., 2022, p. v).

One of the responses to these alarming shortcomings was, for example, *The Vienna Manifesto on Digital Humanism,* which was developed in May 2019 by a group of more than 20 scholars in different fields. Through this manifesto, they urged academics, industrial leaders, professionals, and politicians to become actively involved in policy formation around technological development—given that it is shaping the world as we know it. As Werthner et al. (2022) expose,

the flood of data, algorithms, and computational power is disrupting the very fabric of society by changing human interactions, societal institutions, economies, and political structures...This disruption simultaneously creates and threatens jobs, produces and destroys wealth, and improves and damage our ecology. It shifts power structures, thereby blurring the human and the machine. (p. xi)

It is important that we develop and design technologies in accordance with human needs and values, rather than allowing our technologies to shape human beings. This is what digital humanism is about. The core principles of the *Manifesto on Digital Humanism* are oriented towards this goal. Some examples of these principles are: "digital technologies should be designed to promote democracy and inclusion," and "decisions with consequences that have the potential to affect individual or collective human rights must continue to be made by humans" (Werthner et al., 2022, p. xii).

In Chap. 2 of this book, we argued for the necessity of overcoming the researcher-practitioner binary in the field of CIE. An example highlighting this distinction between practical and theoretical roles is presented by Dolby and Rahman (2008), who contended that "comparativists tend to focus on academic policy research largely removed from the questions of context and application, whereas internationalists are more concerned with the specific context, location, and application of their research" (p. 680). We maintain this distinction is not entirely real, as scholars in the field often engage in both actions, integrating practice with research and vice versa. Nevertheless, when considering the risks and benefits of incorporating AI tools in the field of CIE, distinguishing between research and practice can provide valuable insights. When examining research, we explore the implementation of AI at the methodological level (explored more extensively in Chap. 5). Meanwhile, when contemplating the implementation of AI in CIE practice, we investigate the various AI tools practitioners employ. The following sections delve into the ethical implications of applying AI to both research and practice within the field. Given the ongoing development, the impacts on research and practice cannot be accurately predicted. Hence, this chapter primarily focuses on offering caution and highlighting present and potential advantages. As in many positivist and uncritical fields, the benefits of implementing AI tend to be more evident than the risks, it is crucial to emphasize the need for caution in navigating the evolving landscape of AI.

Cautions for Using AI in CIE Research

The first caution CIE must consider when using AI for research is its potential to accentuate disparities among worldwide researchers with uneven access to technologies, AI tools, and training, among others. This is likely to worsen the dominance of Western research, a problem prevalent in the field before the advent of AI. Even if researchers from the West include countries from the Global South in their studies, we still talk about the researcher being an outsider. If the CIE field aims to enhance diversity in researchers, AI might represent an obstacle and signify a step back in the efforts to decolonize CIE scholarship and the field. Ultimately, research projects compete for funding and publication, with the advantage increasingly tilting toward researchers with access to superior technology capable of producing previously unimaginable results. While research using AI technologies and tools may contribute to a deeper understanding of phenomena and new discoveries, there are ethical implications when it comes to who is producing this knowledge and from where. Is more knowledge always the best option?

Additionally, introducing AI into education can exacerbate the positivist approach that already dominates the field, as highlighted by Saltman (2022), who notes that "perhaps more than any other field, the imperative for positivism pervades education. Public schools use test-based accountability, in which learning equates with numerical test scores, and changes to teaching and administrative practices are to be guided by numerical outcomes" (p. 1). The main consequence of this positivist approach guiding educational practices is that reforms in teaching and administrative practices are often guided solely by numerical outcomes, overlooking the inherent complexities of education, its diverse learners, and local contexts. In other words, the positivist approach of AI in CIE can hinder the creativity of researchers in the field when it comes to thinking about better and more just futures. This mechanistic approach may further undermine the role and importance of teachers and schools. Instead of prioritizing investments in AI, UNESCO Miao et al. (2021) emphasizes the urgency of directing resources towards schools and teachers. Shockingly, with 244 million children and youth still out of school, the global community continues to underfund the institutions that can effectively address the educational challenges (Miao et al., 2021). The risk lies in neglecting the solutions provided by teachers and schools in favor of the allure of technological solutions, potentially exacerbating the positivist tendencies ingrained in the educational system.

The incorporation of AI in CIE also has the potential to aggravate biases, primarily because it is a relatively new field that has predominantly developed within and for nations of the Global North. The limited diversity among its designers and their ontologies and epistemologies introduces a significant risk of amplifying existing biases within the realm of CIE through the utilization of AI. Such a scenario can contribute to the reinforcement of divisions and discriminatory practices. AI systems must be meticulously designed collaboratively to ensure fairness and equity, actively avoiding perpetuating or amplifying pre-existing biases. However, CIE does not have control over the development of these tools. The biases ingrained in AI design are a direct consequence of digital coloniality. Researchers like Bon et al. (2022) and Mohamed et al. (2020) have uncovered discriminatory biases in AI algorithms, such as algorithms that whiten Asian and Black faces and have trouble recognizing Black faces. As Bon et al. (2022) assert, "these trivial examples show biases embedded in apparently value-free technology, in which existing patterns are unconsciously replicated. These biases pop us unexpectedly in autonomous smart systems and may intentionally or unintentionally exacerbate inequalities" (p. 64).

Finally, in the realm of CIE, the incorporation of AI tools raises critical concerns surrounding privacy and consent. Researchers employing AI must adopt rigorous measures to safeguard the confidentiality and security of individuals' data, including ensuring the data is anonymized and secure from unauthorized access. Ensuring informed consent becomes a pivotal step in navigating the ethical complexities of AI usage. Within CIE, this introduces an additional layer of risk that researchers must consider, as the field of AI does not have the same understanding of informed consent and privacy as CIE and the educational field in general have. During the early development of AI, most research involving human data did not require consent from the individuals the dataset represented. This is because the way it was used in the fields that would initially involve AI (like mathematics, statistics, and computer science) posed little to no risks for human subjects (Crawford, 2021). Given that Institutional Review Boards (IRB) have maintained this position for years, the protection and consent of human subjects are still underdeveloped in the field of AI. Although potential harms expanded since AI moved beyond the initial laboratory-based disciplines, "there is still a strong assumption that publicly available data sets pose minimal risks and therefore should be exempt from ethics review" (Crawford, 2021, p. 116). The dangers involving research in AI continue to grow as researchers have increasing access to data gathered with no or minimal consent and that does not require any form of interaction with subjects. This abstract and impersonal relationship between the researcher and subjects increases the risk of decontextualizing experiences. An example of the harm this uncontrolled access to datasets can have on individuals is the proliferation of gang data used by the police. These datasets defined by predictive AI systems have been shown to be quite inaccurate and biased, with an overrepresentation of Black and Latinx people (Crawford, 2021). Balancing the potential benefits of AI-driven insights with the imperative to uphold individual privacy and consent becomes central for fostering responsible and ethical research within CIE.

Cautions for Using AI in CIE Practice

Again, inequality is also a significant issue for CIE practice, given the unequal distribution and access to AI tools. For instance, UNESCO (2023a) disclosed that it would currently cost USD 1 billion per day to maintain connectivity for education in developing countries. Despite efforts by organizations like the World Bank to enhance connectivity access in Africa to achieve universal connectivity by 2030, the current situation is far from ideal. Internet access exhibits substantial variation among countries, creating a digital divide that mirrors existing educational inequalities. For instance, as of 2021, nearly 70 percent of the population in Ghana had connectivity, while only 8 percent of individuals in South Sudan had access to the internet. The transformative power of AI in education is contingent upon internet access, and the stark contrast in connectivity levels implies that AI interventions may unintentionally widen the educational gap between industrialized countries and emerging markets and between high and low socioeconomic classes within nations. Despite the growing trend of individuals connecting to the internet, countries on the periphery of capitalism still lack digital inclusion and suffer from the problem of digital illiteracy. The disparity in internet access becomes a critical factor that hinders the potential of AI to be a force for educational equity on a global scale. Therefore, CIE practitioners must question whether they are widening the educational gap by enhancing the use of AI in educational settings and focus more on finding solutions to AI's uneven accessibility before recommending the widespread application of AI in educational contexts. Does the availability of AI tools alone necessarily signify it is the most suitable option for education in our current times?

Inequality is not the only problem that CIE practitioners must consider when using AI. The only way for AI to really advance in the education sector is through the platformization of education and the live collection of Big Data in schools (Filgueiras, 2023). This, however, comes with numerous risks and cannot be implemented without a thorough process of ethical planning and convergence throughout and between nation-states. These risks include privacy concerns, algorithmic biases and injustices, behavioral changes in people (including compromising the idea of citizens), and governance challenges in education. The foremost concerns are introducing AI into classroom settings and safeguarding students' privacy. The utilization of AI is inherently tied to platforms that continuously gather data about users, monitoring their every move. For CIE, the pivotal question becomes how to ensure that the integration of AI in education does not compromise users' privacy. As outlined in the discussions of privacy issues in CIE research, unchecked access to user information can give rise to biases and discriminatory outcomes. The challenge lies in establishing mechanisms that prevent the undue exposure of user data addressing privacy concerns to mitigate the potential risks associated with AI implementation in educational settings.

Stemming from the design flaws of AI technology, algorithmic biases and injustices arise when machines make erroneous decisions due to inconsistent data or incorrect supervision during learning processes. This results in the creation of new forms of discrimination and inequalities. For instance, AI, predominantly designed and developed by countries in the Global North and mostly by white individuals, may suggest inappropriate content and erroneously determine what students should learn and when. This compromises the educational experience and reinforces existing biases and disparities within the system. Another pressing concern in integrating AI for CIE is the behavioral impact on individuals. The issue stems from data privatization, allowing AI to manipulate user behavior by tailoring content delivery. For example, adopting face recognition as a security measure can lead to profiling and discrimination. The personalized experiences offered by AI contribute to a depersonalization of identities and generate divides. AI also compromises the idea of citizen because, as AI becomes more widespread, the digital world becomes increasingly dominated by autonomous systems that make the decisions for their users, inevitably impacting the user's autonomy and privacy. As Inverardi (2022) explained, "in a digital society where the relationship between citizens and machines in uneven, moral values like individuality and responsibility are at risk" (p. 26).

Finally, governance challenges, particularly data governance, form another critical risk in implementing AI within education. As platforms continually collect data without users' knowledge or with limited understanding, this constant surveillance alters power dynamics, granting substantial influence to private companies and governments that receive this data. The challenge of AI's potential to influence individuals and elicit particular behaviors underscores the importance of implementing strong policies and effective regulatory measures. These safeguards aim to protect user data from misuse and uphold ethical standards in the continually evolving field of AI in education. This is increasingly important for the field of CIE as it continues to employ AI in its practices. How can the field ensure it does not promote surveillance from big companies and organizations and enhance their control over education systems? This significant concern is explored more deeply in Chap. 6 of this book.

ADVANTAGES OF IMPLEMENTING AI IN CIE RESEARCH

Despite all the cautions aforementioned that the field of CIE must consider when integrating AI, there are numerous evident advantages to using AI in both CIE research and practice. Implementing AI in educational research offers a significant advantage in increasing efficiency and speeding up the development of studies. Basically, AI can function as a co-researcher by aiding with specific tasks and processes that would otherwise require a lot of time and resources. A recent example in the field of science is Google's creation of the AI model *Gemini*,[2] a tool capable of automatically updating scientific literature on various topics and even generating

---

[2] For more information about Gemini explore Google's website: https://blog.google/technology/ai/googlegemini-ai/#sundar-note

graphs. The conventional scientific literature review is time-consuming for researchers who manually seek and extract data from a wide array of scientific sources. *Gemini* streamlines this process, requiring researchers to grant data access only. The application then retrieves all relevant information within minutes. This example highlights how AI aids researchers in optimizing their resources, enabling them to allocate more time to analyzing and applying their findings. The same principle holds for the broader field of education and, more specifically, CIE. AI's rapid data processing capabilities enable large-scale comparative analyses across diverse educational contexts. Researchers can efficiently analyze trends, identify patterns, and draw insights from extensive datasets, facilitating evidence-based decision-making and contributing to the continuous improvement of educational practices worldwide.

Access to a more diverse range of data is another crucial advantage AI facilitates in educational research, potentially promoting inclusivity and global collaboration. Overcoming geographical and institutional barriers, AI provides researchers access to a wider variety of sources, enhancing the comprehensiveness of their findings. Additionally, AI tools facilitate the collaboration between researchers or participants from different parts of the world, fostering partnerships that are often essential for addressing complex educational challenges. Access to diverse data, however, is contingent on having a field of AI that includes developers from different parts of the world, especially from marginalized nations.

A third advantage is creativity, enhancing new ways of thinking about the future. The collaboration between humans and machines introduces new possibilities in educational research because it is a new form of collaboration. As described by Dr. Maynard in an interview with Richardson et al. (2023), AI contributes to expanding our perspectives on the future through co-creation and co-thinking with machines. Although the current development of AI is limited to a narrow segment of the global population, the potential for creative opportunities would significantly increase as AI development and design become more inclusive and representative. AI also offers the possibility to minimize bias in data analysis, but only if the design of AI systems is approached collaboratively. By incorporating diverse perspectives and ensuring inclusive training, AI can contribute to more objective and unbiased analyses of educational data—as machines have the potential to operate from a multiplicity of viewpoints and could lack a specific positionality.

Finally, breaking down language barriers in academia is another advantage of AI, enabling data collection and presentation of results across different languages. AI's language translation capabilities facilitate communication and collaboration on a global scale, allowing researchers to access information and contribute to the academic discourse regardless of linguistic differences. This enhances the inclusivity and accessibility of research, fostering a more interconnected and diverse scholarly community. This is especially relevant in the field of CIE, given that it is an international field that encourages collaboration across nations with diverse languages. AI's translation capabilities could also help the field solve researchers' unequal access to academic publishing because of language barriers, leveling the playing field.

ADVANTAGES OF IMPLEMENTING AI FOR CIE PRACTITIONERS

Implementing AI in educational practices extends to various planes, including alleviating teachers' workload and enhancing school management capabilities. For instance, AI-driven tools can assist educators in reading and grading students' assignments, saving valuable time that can be redirected towards personalized student interactions, lesson planning, or professional development. This contributes to reducing teacher stress and fosters a more supportive and conducive learning environment. Regarding school management, integrating Big Data and AI allows for real-time monitoring and analysis of various educational parameters. By operating on a platform model with constant data collection, schools can develop and employ autonomous decision systems to help them use resources better, make administrative tasks easier, and become more efficient overall. For example, AI-driven systems can automate administrative tasks like scheduling, grading, and attendance tracking, freeing up time for educators to concentrate on instructional design and student engagement strategies.

AI can also assist CIE practitioners in increasing accessibility and expanding educational opportunities. AI tools can support integrating learners with diverse needs and disabilities by providing adaptive learning environments tailored to individual requirements. Language barriers can be overcome through AI-powered translation services, making educational materials accessible to students from different linguistic backgrounds. Additionally, AI enables the implementation of more flexible and individualized education approaches. For instance, adaptive learning platforms can dynamically adjust the curriculum based on each student's progress, learning style, and preferences. This enhances the overall inclusivity of education and fosters a personalized learning experience for students with varied needs and backgrounds.

CONCLUSION

In this chapter, we have examined the issue of digital coloniality and emphasized the imperative for CIE to actively avoid perpetuating it, particularly in light of the growing presence of AI in the field. In summary of this analysis, we argued that the advancement of decolonizing efforts within CIE during the era of AI brings two main challenges to the forefront: access issues and the explosive origins of AI tools. The ethical dilemma of embracing AI opportunities in a global landscape characterized by inequality raises essential questions: *Is embracing AI opportunities ethical in a world with unequal access to resources and technologies and, therefore, uneven possibilities to benefit from AI? Is CIE, through its employment of AI, an active accomplice in perpetuating educational inequalities? How can the field benefit from the advantages of AI while at the same time questioning and offsetting its risks?*

The issue of who designs AI tools becomes a crucial consideration for CIE scholars striving to enhance education systems that challenge coloniality and promote decolonial practices. Within this challenge, Lee (2022) provocatively argues that rectifying ethical concerns surrounding AI requires a departure from "digital creationism"—the belief that harmful outcomes generated by digital technologies emerge from unethical actions of the creators of these technologies, who only follow profit and have limited regulations—and a shift towards "digital coevolution." According to Lee (2022), this latter paradigm underscores the shared responsibility of both users and creators in addressing the ethical dimensions of AI. This paradigm shift assigns a significant responsibility to education, as "it may be just as effective to pass laws that focus on educating the public, for example, as it is to pass laws that regulate the technology producers" (Lee, 2022, p. 6). Within this realm, CIE emerges as a pivotal player tasked with researching and expanding educational practices that ensure AI's ethical and equitable use on a global scale.

On a hopeful note, as we have highlighted in the exploration of the different advantages AI tools can provide, AI holds the promise of unlocking unprecedented opportunities for both researchers and practitioners in the field of CIE when employed judiciously and guided by clear ethical frameworks. The potential to accelerate processes, enhance accuracy and efficiency, and acquire previously unattainable data highlights AI's transformative impact on advancing the field and providing valuable tools for improving education systems. As UNESCO (2023a) emphasizes, prioritizing research and learning outcomes over mere digital inputs is critical. In other words, to help improve research and implementation of educational strategies, digital technology should not be a substitute for but a complement to face-to-face (human) strategies. We must learn to live both with and without AI, which encompasses the field of CIE.

In navigating the intersection of CIE and AI, the imperative extends beyond confronting challenges to actively shaping a more inclusive and ethically grounded educational landscape. Comparative studies can be instrumental in fostering a more critical approach to AI's impact, offering insights that contribute to more informed decisions within educational practices. The call for CIE scholars to engage in comparative analyses, scrutinizing how AI affects various fields across diverse countries, positions them as key contributors in providing recommendations, suggestions, and warnings in the ever-evolving landscape of AI in education. With this goal in mind, fostering cross-collaboration among scholars from different fields becomes crucial, allowing for a more comprehensive mapping of AI's impact. In essence, the role of CIE in the era of AI goes beyond addressing challenges—it involves actively and proactively shaping a future where education is not only technologically advanced but also ethically guided and universally inclusive.

CHAPTER 5

Using Artificial Intelligence for Educational Research: Methodological Implications

Abstract This chapter will explore the methodological implications of AI in CIE research and the efforts to decolonize research methods used in the field. The conversation on AI's role in decolonizing CIE research continues to play out in real-time. However, even as these ethical questions abound, it is important to highlight some of the helpful applications of the myriad new tools and techniques that AI has introduced for research and scholarship purposes. AI-powered conversation robots, machine learning, natural language processing, and predictive analytics tools offer exciting opportunities to assist CIE researchers with their work. Therefore, this chapter will also explore how AI is being used to benefit researchers as they navigate the often lengthy, multi-step process of collecting data, writing, and publishing studies. The chapter will conclude with some recommendations for ethical research practices using AI in CIE.

Keywords AI in research • Research methodologies • Decolonial research methods

INTRODUCTION

As noted in previous chapters, researchers in CIE study the world's educational systems to discern what can be learned from approaches in global educational contexts and how these approaches might apply to other educational systems. As technology has continued to evolve throughout the last few decades, research has shifted alongside it with breakthroughs in telecommunication. Online publication of journal articles and using e-mail, web pages, and computer listservs facilitate rapid communication between comparativists and disseminate the field's trends more widely. With the rise of search engines and the reduction in the price of low-cost computing devices, ranging from handhelds to mobile phone to laptops or netbooks, CIE research became more accessible worldwide. The emergence of AI is yet another progression in these technological developments. As Chen et al. (2020) note

> AI initially took the form of computer and computer related technologies, transitioning to web-based and online intelligent systems, and intimately with the use of embedded computer systems, together with other technologies, the use of humanoid robots, and web-based chatbots. (p. 75264)

Chapters 1 and 2 discussed how researchers have used AI in CIE to help them collect, sort, analyze, and interpret data. The term data is typically used in the technical disciplines to mean content that is "efficient for 'data mining' techniques [and] for classifying, quantifying, and extracting useful information" (Nilsson, 2009, p. 111). However, with the emergence of AI intermingling with everyday research practices, the very notion of what "data" means may need to be reconstituted into CIE scholars' working knowledge if the field is to succeed in its aim to decolonize its research products and processes.

Many CIE scholars have explored the potential of digital data to effect change in comparative methodologies of education research (Salajan & jules, 2020; Görur et al., 2018). However, not all change is just. Salajan and jules (2020) elucidated some of the dangers of data use in the educational intelligent economy

> The continuous challenges in regulating data flows to discourage illegitimate or harmful uses through responsive policy mechanisms and instruments present a complex problem for educational researchers in the educational intelligent economy, as the sophisticated machinery of data production continues to accelerate its pace to advance public interests, but, more frequently, serve bankable intents. (p. 5)

Responsive, rather than proactive, policymaking has allowed the data mining and AI systems manufacturing industries to function with little regulatory oversight. The "bankable intents," or the prioritization of profits for the actors behind these developments, are also functioning to proliferate the neoliberal priorities of the Fourth Industrial Revolution. As discussed in Chap. 1, scholars and researchers in CIE working within the context of rapid technological change and economic value-centricity are not new. The advent of AI has been a game-changing technological innovation for research and pedagogy in CIE by training machines or computers to execute tasks that would typically require human intelligence, such as learning, problem-solving, and decision-making (UNESCO, 2019a). However, "there has been a general failure to address how the instruments of knowledge in AI reflect and serve the incentives of a wider extractive economy" (Crawford, 2021, p. 135). To examine how AI can perpetuate colonial practices, Chap. 4 highlighted the environmental peril brought on by the actions of humans in the current Anthropocene era and the socioeconomic stratification exacerbated by the use of data as capital. It also illuminated how, in the context of environmental chaos, the exploitative nature of capitalism and neoliberalism, and the ongoing effects of colonialism and imperialism, AI has the potential to exacerbate the already vast levels of social inequality between the Global North and the Global South (Nordtveit & Nordtveit, 2020). Quite literally, fossil fuels used to power data mining by companies in the Global North are furthering the catastrophic ramifications of climate change for communities in the Global South. This social stratification and the anthropocentric positioning of data as a "natural resource" that can be extracted from human beings, just as oil is extracted from the Earth, is clearly "in keeping with the broader neoliberal visions of markets as the primary forms of organizing value" (Crawford, 2021, p. 113). Thus, the emerging ubiquity of AI presents a new challenge for CIE researchers, all while the field is in the throes of epistemic and ontological change, specifically regarding efforts toward its decolonization.

The predominance of industrial data mining to power AI is also in keeping with the prevailing practices contributing to global economic disparities. Mining data from vulnerable populations for the use and benefit of wealthy, male-dominated corporations reflects a continuation of the dominant colonizing attitude whereby indigenous resources, in this case, an individual's image, likeness, ideas, perspectives, creations, and personal data are there for the taking. Thus, poor people can become the targets of the most harmful forms of data surveillance (Crawford, 2021). The unequal levels of global economic development also impact how AI is used in learning, teaching, and research in educational settings in emerging and frontier markets. These issues point to why AI tools can function as a stumbling block to the ongoing effort to decolonize research in CIE.

One of the primary questions this book seeks to explore is, *What has been the role of education and technology in helping humanity think before innovating?* Previous chapters up to this point have recounted the history behind the rise of AI and AIED, considered the benefits and areas for caution that AI technology poses, and discussed the ways that AI may disrupt efforts to decolonize the field of CIE. Building on this foundation, this chapter will explore the methodological implications of AI in CIE research and the efforts to decolonize research methods used in the field. The conversation on AI's role in decolonizing CIE research continues to play out in real time. However, even as these ethical questions abound, it is essential to highlight some helpful applications of the myriad new tools and techniques AI has introduced for research and scholarship purposes. AI-powered conversation robots, machine learning, natural language processing, and predictive analytics tools offer exciting opportunities to assist CIE researchers with their work. Therefore, this chapter will also explore how AI is being used to benefit researchers as they navigate the often lengthy, multi-step process of collecting data, writing, and publishing studies. The chapter will conclude with recommendations for ethical research practices using AI in CIE.

Observing Research in Comparative and International Education

Conducting research in CIE requires scholars to have a breadth of knowledge in many interrelated fields and disciplines. CIE scholars utilize a variety of approaches to carry out their work. Within these approaches lies an important difference between methodology and methods, which it is necessary to discuss so they are distinguished early on. Methodology is "the study—the description, the explanation, and the justification—of [the research] methods, and not the methods themselves" (Kaplan, 1964,

p. 18). Methods are "the practical activities of research: sampling, data collection, data management, data analysis, and reporting" (Carter & Little, 2007, p. 1318). This chapter will analyze the implications of AI technologies on CIE methods and methodologies, particularly how AI is connected to ongoing conversations about decolonizing the field's scholarship.

Although they are distinct from each other, comparative education and international education are sister disciplines (Kelly & Altbach, 1986). Klees (2017) articulates the similarities between comparative education and international education and their shared grounding in social science research by stating, "comparison is the essence of science and the field of comparative and international education, like many of the social sciences, has been dominated by quantitative methodological approaches" (para 1). While it is true that scholars in both fields use qualitative, quantitative, and mixed methods, beginning in the late 1960s, quantitative methods were primarily used in CIE in an attempt to broaden the applicability of research findings in the field to other social science arenas like economics and public policy (Klees, 2017). Before this shift toward the field's scientification (see Noah & Eckstein, 1969), comparative education applied qualitative methods as cultural and historical approaches to comparative meaning-making.

Typical audiences for CIE research publications include graduate students, professors, and policymakers. While the two fields comprising CIE share research approaches typical for social sciences, they sometimes contrast in how these audiences utilize research findings. Comparative education research is generally considered a more theoretical field, with international education research being more practitioner-focused (Sylvester, 2003). Dolby and Rahman (2008) argue that "comparativists tend to focus on academic policy research largely removed from the questions of context and application" (p. 680). They and others (see Malakhov, 2020) see comparative research as existing in a more theoretical space that is used to draw contrasts between like and unlike entities to trace the origins of their differences and draw potential conclusions about why those differences exist and what their implications may be for educational systems.

On the other hand, international education is a more applied field wherein scholars and practitioners use research findings to immerse themselves in other cultures to understand them better (Sylvester, 2003). International education also focuses on the practical outcomes of studying foreign systems to practically transmit and foster beliefs, attitudes, and knowledges among nations (Sylvester, 2003). Although these applications can appear to be broad or vague, according to James (2005), Bunnell (2006), and Cambridge and Thompson (2004), the many interpretations and uses of the words "international education" are the primary reason for the uncertainty, confusion, and disagreement regarding what international education is. For the purposes of this chapter, we are choosing to follow the same conclusion reached by other CIE scholars and will characterize the primary function of research in comparative education as concluding contrasts and the primary function of research in international education as identifying practical applications for critical findings in global education settings.

METHODOLOGICAL INCONGRUENCIES IN CIE AND AI

Even as technological conditions advance to assist us, researchers in CIE must continue to navigate ever-expanding bodies of literature and align multiple factors and comparative units in their work. Considering that CIE was developed and currently exists within a colonial paradigm, Görurr et al. (2018) point out that

> certain methodologies have gained traction—among them are randomized controlled trials, meta-analyses, and large-scale standardized assessments which privilege epistemological stances from the natural and statistical sciences; network methodologies which draw from a relational ontology and spatial theories; and digital platforms and Big Data studies and digital methodologies, which are enabled by new and ever-changing technologies. (p. 5)

CIE scholars use these and other methods to develop and interpret statistics and draw conclusions from predictive equations to name and anticipate patterns in data. For example, some quantitative scholars in CIE have conducted statistical analysis of large government-sourced data sets to determine and evaluate educational quality within and between school systems worldwide (Marginson & Mollis, 2001). On the other hand, qualitative scholarship is more concerned with the "hows" and "whys" of individuals' experiences rather than their statistical significance. Another vital feature of CIE studies, especially critical studies such as those that seek to contribute to the decolonization of the field, is the researchers' epistemic lenses or perspectives.

Differing Epistemologies in CIE and AI

At times, differing epistemic perspectives can become so entrenched that scholars working on similar concepts can miss opportunities to make and strengthen the connections and calls to action that are shared between their work. Such discrepancies can obscure the momentum toward transformative research for transformative outcomes in education, which decolonization demands. The intent behind the research itself can also exacerbate epistemic differences between researchers in CIE. For example, research methods that focus on interpreting patterns in data, which are then generalized onto various populations, have a different aim and purpose than methods that center on developing theoretical frameworks specific to the context of where these theories are created and where they will later be applied.

Take, for instance, the theoretical epistemology behind the concept of "decolonization" which can mean different things to different authors and audiences. McDonnell and Regenvanu (2022) define decolonization strictly as the actions of Indigenous communities as they work to regain their rightful land. Their epistemology centers a framing of Land as "essential to ensuring sovereign futures for indigenous people" (McDonnell & Regenvanu, 2022, p. 236). Others use the term "decolonization" when referring to methods of applying various theoretical frameworks and concepts to critique and resist imperialistic systems of domination, repression, surveillance, and dehumanization (Nordtveit & Nordtveit, 2020). This epistemic discrepancy between decolonization as a framework and decolonization as a specific form of action has led to the development of different scholarly narratives in CIE about what it means to "decolonize."

Similar discrepancies are mirrored in the research practices data scientists use to collect data and develop the theories that inform AI. The science used to create AI tools often occurs at a physical and metaphorical distance from the spaces where the tools will be applied. In other words, the scholars developing AI tools often operate from an epistemically incongruent lens compared to the end-users, or the persons who will apply the AI tools in their unique fields. These epistemic differences occur primarily because the academic disciplines that shape AI development, including computer science, data science, statistics, mathematics, biophysics, and more, are rooted in quantitative observations and conceptions of knowledge that claim to produce and identify objective truths under the guise of positivism (Mohamed, 2018; Murphy & Largacha-Martínez,

2022). The algorithmic process of classification that undergirds machine learning necessarily reduces context to make data more computable (Crawford, 2021). This is because AI tools are trained to rely extensively on positivist epistemologies; in turn, they produce results for the end-user based on generalized and applied assumptions without additional contextual nuance.

The epistemological dimension of research is informed by what the researcher values and can heavily shape the tools and methodologies researchers use. Epistemologies are one way for the researchers' beliefs about the meaning and legitimacy of certain forms of knowledge to manifest in their scholarship. Positivism, for example, is an epistemology that holds to a "preoccupation with the instrumental use of knowledge" and is tied to notions of objective neutrality and scientific conclusions as superior to values, feelings, and subjectively defined knowledge (Giroux, 2013, p. 33). Positivist perspectives on research push for the social sciences to be molded against the same assumptions and methods of the natural sciences. This is diametrically opposed to the complex interplay of personal reflection, acknowledgement of power, ideology, and lived experience, which inform critical and decolonial research.

Positivism already dominates the field of CIE, particularly guiding the approaches used to evaluate education. Saltman (2022) claims that "perhaps more than any other field, the imperative for positivism pervades education. Public schools use test-based accountability, in which learning is equated with numerical test scores, and changes to teaching and administrative practices are to be guided by numerical outcomes" (p. 1). The main consequence of this positivist approach to research on educational practices is that reforms in teaching and administrative practices are often determined solely by numerical outcomes, overlooking the inherent complexities of education, its diverse learners, and local contexts.

In contrast to positivist research practices, CIE scholars aim to illuminate the complexity of global education using an interpretivist paradigm. Interpretivism contends that human beings, including the researcher and participants, "are the primary instruments in a study" (Ravitch & Carl, 2020, p. 5). An interpretivist lens centralizes the importance of human context and diversity and considers the unique impacts of these characteristics on the research outcomes. CIE scholars do this by examining educational systems of global populations across broadly defined categories, including spaces, places, times, structures, and cultures. In tandem with the rise of critical pedagogy in the 1970s (see Giroux, 2013), studies in

CIE today also examine social strata including race, class, and gender and how these shape and impact various aspects of global educational systems through policies, curricula, credentialing processes, and more (see Bray et al., 2016). Comparing these units in research requires CIE scholars to develop a deep and contextual knowledge of the factors of comparison. While it is possible to achieve this knowledge by working only with data, the learning process through which scholars' knowledge is developed can also be supported through fieldwork.

Differing Methods in CIE and AI

Fieldwork is essential to qualitative research. It requires "that the researcher is physically present with the people in a community" to engage individuals and generate data from research participants in "settings that are authentic rather than contrived" (Ravitch & Carl, 2020, p. 10). In CIE research, examples of doing fieldwork in "authentic" environments can include conducting sociological analysis of interactions across difference within a given global context (Dolby & Rahman, 2008), or developing historical analyses of the observed impact and change over time caused by various policy decisions in specific community settings (Peters, 2007; Clemens, 2008). These examples point to the meaning and importance of community context during data collection in qualitative research in CIE. However, unlike CIE, which is in line with social sciences research, the pre-cursor disciplines for AI tool development, including applied mathematics, statistics, and computer science, do not share the same emphasis on considerations of context in their methodology. Thus, incongruencies between knowledge production in CIE and the knowledge production powering AI tools can be observed in each field's research process.

The research methods used in CIE, especially those which involve collecting data from human subjects, require oversight and approval from Institutional Review Boards (IRB). However, as was mentioned in Chap. 4, the research that undergirds AI tool development is not typically considered human subjects research and is thus not held to the same ethical standards to mitigate potential harms to participants. Furthermore, AI tools are often developed using publicly available data or data purchased from private companies. Thus, the human subjects from which the data were extracted are not afforded the same ethical protections as those whose data are collected first-hand from researchers in the field. As Crawford (2021) notes, "there is still a strong assumption that publicly available data sets pose minimal risks [to human subjects] and therefore should be exempt from ethics review" (p. 116). However, there are considerable ethical risks regarding the use of AI in research; some of these risks, specifically research bias and privacy violations, were discussed in Chap. 4. Using Big Data as an example to illuminate the risks CIE researchers need to be aware of when using AI in their studies, the following section will overview precautions for data collection, data analysis, data sharing, and data reporting when using AI tools in research.

Precautions for Data Collection with AI. Big Data mining is the process of extracting large volumes of data from various sources at a high speed or velocity (Laney, 2001). The extracted data can include, but are not limited to, biometric data, such as that used by facial recognition algorithms and speech authenticators; photos, such as those posted online; and electronically published written content (Crawford, 2021). These forms of data can be directly or indirectly collected from human beings and used to train AI machines, while the companies that acquire the data via the mining process often skirt the need for agreements or signed releases. Unlike in academic settings where researchers are required to obtain informed consent from participants when collecting data from them as part of the research ethics review process, electronic "terms and conditions" documents are often used in the data mining industry to constitute informed consent. This practice facilitates massive data collection without users' mindful awareness (Crawford, 2021). One precaution researchers using Big Data should consider is to report on the process they followed to obtain the data used in their study and where it was acquired. Appending or referencing the data privacy policy of the source for the data and for any AI tools utilized in the data analysis process may be another burgeoning best practice.

Precautions for Data Analysis with AI. Chapter 4 provided some examples of the ethical repercussions of the "rapacious international culture of data harvesting that can be exploitative and invasive and can produce lasting forms of harm" (Crawford, 2021, p. 118). However, much of the harm that can be produced in research settings occurs after the data are collected in the analysis stage. Just as CIE researchers can inadvertently draw biased conclusions from the data they collect, so can the researchers and data scientists that inadvertently train AI tools to produce biased conclusions. As Crawford (2021) points out, "every dataset used to train machine learning systems, whether in the context of supervised or unsupervised machine learning, whether seen to be technically biased or not, contains a worldview" (p. 135). And its worldview will likely be imbued with the creator's biases.

There are generally two types of bias that AI tools may exhibit that CIE scholars using them or using Big Data may need to know. The first is statistical bias, which is not much different from the same concept in quantitative research, where it refers to systematic differences between a sample population and the actual population, making the sample not truly reflective of the whole (Mitchell, 1980). The second form of bias, machine learning bias, refers to a type of error that can occur during the predictive process of generalization, namely a persistent classification error that the system exhibits when presented with new examples (Crawford, 2021). This can produce faulty results when the AI tool fails to classify data accurately in accordance with how it was trained (e.g., incorrectly predicting patterns in language or failing to distinguish between two similar images based on their significant differences, such as color, shape, etc.). Currently, there are minimal means for end-users, such as CIE researchers, to contest the instances of bias that may occur when AI tools are used to expedite data analysis. One potential best practice could be to reference or include information about the validity and reliability of the AI tool used, when available, or to reference the tools' success rate when making classifications in its training data set(s). These suggestions for transparency point to data reporting as another area of caution for researchers using Big Data or other AI tools in their work.

Precautions for Data Sharing with AI. The importance of data sharing or "making the data which underpins empirical research papers available to the researcher community so that others can undertake their own analyses and build upon it for further research" (Rushby, 2013, p. 676) is echoed throughout academic disciplines as essential to researcher accountability and to the advancement of new knowledge. Academic organizations, from the Committee on National Statistics to the American Educational Research Association (AERA), have published numerous reports urging the importance of data sharing in their fields, regardless of the methodology applied or whether the data were accessed online or generated through fieldwork. Furthermore, some academic organizations such as AERA are now funding data mining projects themselves to identify pertinent new research questions for which they may choose to fund projects. This process is greatly assisted by AI, which can help these organizations and their partners identify patterns in vast amounts of Big Data.

However, Nordtveit and Nordtveit (2020) argue that "data sharing continues to reinforce imperialism through control, dissemination, and application of data, and how electronic and digital colonialism preserve current intellectual and structural hegemonies" (p. 33). They connect this claim back to the inequitable power of academic funders to shape future research directions based on what they deem profitable. Thus, the widespread requirements for data sharing, which often is requested of the author at the time of publication, can not only expose research participants to the privacy concerns mentioned before—for instance, with these data sharing requirements in place, how can scholars ensure the anonymity of interviewees in qualitative studies?—but can ultimately make researchers beholden to "certain methodologies" (Görur et al., 2018), which generate the most data for potential profitability, as mentioned earlier in this chapter. Given that data sharing is a requirement in many fields now, all researchers may be exposed to these impacts, whether they use AI tools in their studies or not. However, with this knowledge, they can make an empowered decision on what methodologies to use. A best practice may be for more scholars, regardless of their discipline, to consider adding a section to their papers justifying their methods rather than simply describing them. If more researchers provide a methodological rationale while writing and present that alongside their findings, the data sets they share in the publication stage could become more contextualized and thus retain some of the meaning lost in the process of mining Big Data.

Precautions for Data Reporting with AI. Researchers already have several important decisions to make as they prepare their work for the final steps before publication. In addition to the extra work needed to prepare their data to be shared alongside their published study, there is also now the new ethical question of when, whether, and how to give credit to AI tools for the role they may have played in the research process. For example, it is common for authors to provide examples of search teams used in databases to find articles in a literature review. Will authors using AI need to have a standardized way of explaining the "prompts" or other human inputs used to get to any AI outputs used in the research? Because of the powerful and various ways AI might be used, it is becoming increasingly important to note the use of AI in research outputs or at least describe it in the final research reporting product. The rationale for disclosing the use of AI tools in research may be the same as the rationale for data sharing: to increase the reliability and replicability of empirical research findings and codify them as valid sources of new knowledge. However, in the process of disclosure, new ethical questions will undoubtedly emerge about the legitimacy of AI-assisted work, especially when considering the vastly uneven access to AI tools between scholars in the Global North and those in the Global South. Regardless, a burgeoning best practice for scholars using AI may be to disclose what tools were used in the study and how, if not for the advancement of academic knowledge, then for the advancement of ethical use of AI in different fields by least beginning to track *how* it is being used. Unfortunately, "once AI moved out of the laboratory contexts of the 1980s and 1990s and into real-world situations the potential harms expanded" (Crawford, 2021, p. 115). Therefore, scholars across disciplines may find that they have an ethical duty to share with their peers how AI has shaped and informed their work.

Despite these and many other concerning ethical ramifications, AI can be helpful to CIE scholars as they work to generate new research. Whether through conversation robots, natural language processors, or predictive editing tools, these machines can function as a human-like partner to assist researchers in CIE. Some examples of AI tools' research functions include synthesizing data quickly, recalling and communicating historical facts, expediting reviews of the literature, and assisting with grant writing with helpful predictive prompts. CIE scholars can also use AI to augment or accelerate our research by decreasing the amount of time and cognitive energy needed to complete the tasks of scholarship, thus helping researchers increase their productivity. With its widespread dissemination capacity, AI is also transforming the potential applications of CIE scholars' findings. Bearing the myriad applications and innovations of AI in mind, the following section will describe just some computer programs and algorithms that have been trained to "think humanly" to assist scholars in CIE research. Table 5.1 outlines some of the AI tools that are commonly used in research today.

Table 5.1 Selected AI research tools

Tool name	Research purpose	Tool developer	Launch year
Elicit	Literature reviews	Ought	2017
Consensus	Literature reviews	Consensus	2022
GPT-4	Data analysis and visualization	Open AI	2023
Grammarly	Text editing (spelling and grammar)	Grammarly	2009
ChatGPT	Text editing (reference lists)	Open AI	2021

APPLICATIONS OF AI IN CIE RESEARCH

As a reminder, in this book, AI is generally defined as the ability of machines or computers to execute commands and tasks that would typically require human intelligence, such as learning, problem-solving, and decision-making. AI is good at interpolation, in other words, helping researchers with connecting the dots. The following sections will briefly overview five different AI tools that are currently on the market and will outline the ways they can assist researchers according to information on the tool developers' websites.

AI for Literature Reviews

AI tools may provide researchers with access to more data from lesser-known sources, which may help amplify perspectives and knowledges from underrepresented communities become more included in the research process. By combing through millions of online publications to analyze texts for keyword patterns, AI platforms such as Elicit and Consensus are being used to assist in the research process even before scholars begin the process of data collection. Ought, the company that developed Elicit, boasts that their tool can "analyze research papers at superhuman speed." Elicit users type in a research question, similar to how one would in ChatGPT, and the tool produces a list of relevant papers from a database of 200 million publications and provides a one-sentence abstract summary for each. Researchers can then group these results into appropriate categories of their choosing and Elicit will extract details from the papers into an organized table. These functions significantly reduce the time needed to conduct literature reviews and could virtually eliminate the time it takes to produce annotated bibliographies. Consensus, the AI tool by the developer of the same name, also assists researchers with identifying studies for literature reviews and uses large language modelling techniques to summarize the results.

Another development in AI that holds potential for CIE is "literature-based discovery (LBD)" which involves analyzing the existing topical literature, using language analysis tools like Elicit and Consensus to look for new hypotheses, connections, or ideas that scholars may not have explored yet (OECD, 2023). According to the OECD (2023), what sets LBD apart is that while other tools "use statistics and 'interestingness' measures to identify explicitly stated findings or significant associative trends in the data, LBD attempts to identify unknown knowledge that is implicitly rather than explicitly stated" to help researchers develop new hypotheses based on keyword combinations that are *absent* from the current data available (p. 141). Thus, with the aid of AI, CIE scholars can expose less-examined aspects of the literature and develop wholly new hypotheses using data from different communities in new ways.

AI for Data Analysis and Visualization

Another way CIE researchers can deploy AI to augment their cognitive capacity is by using software to identify promising patterns in data for analysis. One example of an AI-powered data analysis software is GPT-4, a new deep learning product from OpenAI, the maker of the popular tool ChatGPT. According to the developers, "GPT-4 is a large multimodal model that, while less capable than humans in many real-world scenarios, exhibits human-level performance on various professional and academic benchmarks. For example, it passes a simulated bar exam with a score around the top 10% of test takers." GPT-4 accepts image and text inputs and emits text outputs, as well as data visualizations like charts and maps, which can be imported via PDF text files. As Nilsson (2009) states, "machine learning methods are playing an increasingly important role in data analysis because they can deal with massive amounts of data. In fact, the more data the better" (p. 111). Thus, the value-add of tools like GPT-4 for researchers is that it can analyze and create visualizations from much larger data sets much more quickly.

AI for Text Editing

Finally, one of the most commonly used and increasingly taken-for-granted forms of AI assistance are the different text editing tools which many of us already use. The spell-check features in rich text tools that are utilized for word processing and email, along with the predictive text features that can "fill-in-the-blank" for sentences in progress, are different examples of natural language processing tools. Millions, including researchers, use the now familiar platforms Grammarly and Chat GPT to expedite the writing process. In addition, speech-to-text technologies, such as the auto transcription feature in the Zoom video conferencing platform, are trained to detect languages using vocal patterns and then translate the audio into typed words.

IMPLICATIONS OF AI FOR DECOLONIZING CIE RESEARCH

When Nordtveit and Nordtveit (2020) explored the educational intelligent economy using a decolonial framework, the authors argue that while technological advancements are not necessarily a bad thing in and of themselves, there continues to be an urgent need to investigate how technology can be developed, shared, and sustained to create a more just, inclusive, and transparent, in other words, decolonial, system of research and academic inquiry. This can be achieved if not only the process of academic research is more community-centered, but also if the terms of the research engagement become more directly community-led (Curtis, 2023; Nordtveit & Nordtveit, 2020). In other words, unless we are accountable to, collaborating with, and reliant upon community members—not institutions, funders, or industries—in the process of research, we cannot begin to disrupt colonial research practices in CIE. In addition to Nordtveit and Nordtveit (2020) and Salajan and jules (2020), many other scholars in CIE have produced conceptual work that argues for the field's decolonization and urges other CIE researchers to adopt decolonial approaches. Cortina (2019) echoes the need for this work in her article on disrupting North-South directionality. She invites researchers to join her in this endeavor by conducting studies that

> stop the destruction of Indigenous knowledge that is taking place through imposition of the dominant epistemic regime and encourages instead the development of decolonial thinking and research paradigms that contest North-South hierarchies in knowledge in order to promote equality and justice in local and global communities. (Cortina, 2019, p. 463)

Keita Takayama has also published numerous studies on decolonial and postcolonial scholarship as intellectual resources for CIE scholars to tackle active colonial legacies. Specifically, they focus on applications of Southern Theory (see Connell, 2007) in CIE to disrupt North-South domination in academia. Takayama's (2022) recent article on "doing Southern Theory" from a Shinto Japanese perspective is a meaningful, critical reflection on their positionality and how this informs their lens on decolonial research even as they continue to be shaped and changed by encounters with other critical theories, such as ecofeminism, that challenged them to reforge their perspectives. This practice of critical reflection and integration of new ideas is essential to decolonial praxis, and they modeled it well. Yet in an earlier publication, Takayama (2020) also acknowledged the limitations of the hyper-conceptual scope of CIE research, and how this can limit the actionable impacts of the field's decolonial scholarship. They note "the lack of methodological discussion in the field of comparative education about the very political and epistemological tensions [it explores]," though they speculate that "this omission has to do with the fact that most of the researchers in the field do not publish their research in the language of the targeted countries" (Takayama, 2020, p. 61). Still, Takayama (2020) encourages CIE scholars "to think deeply about the potential limitations and dangers of the decolonial knowledge project, for instance, that focuses rather exclusively on relativizing and denaturalizing the assumed universality of Western modernity and its theoretical projects" (p. 61). This example is a perfect metaphor for why AI tools cannot be a panacea for all of the issues in CIE as the field works to decolonize itself; for while many AI platforms can mass-translate research publications into other languages, there are not AI tools that can replace the social, political, and epistemological perspectives and will of CIE scholars who are working to disrupting colonial practices in the field. Building the political will and skills necessary for actionable decolonial research is a project that can only be enacted through community engagement and commitment.

Throughout this book, we have echoed Takayama's (2020) call to action for comparative education research

> to hold itself accountable for the political and epistemological consequences of its own research not simply at the level of international scholarly discourse where an anthropological approach to cultural relativism is valued but also at the level of domestic (national) political struggles where the internal coloniality of power operates to create serious material consequences on the basis of various "differences." (p. 61)

Answering this call requires CIE scholars to conduct rigorous studies on the ground and in the field with community partners leading the way and in proximity with those who experience the "serious material consequences" of imperialism, including digital imperialism, which is exacerbated by the process of AI development and use in research. While we have not yet identified any empirical studies in CIE that have integrated decolonial methods with a critique or examination of AI in a community context, there are many other scholars within and beyond CIE doing decolonial, community-centered work which can provide a framework for future studies to examine the impact of AI.

Decolonial Research Method Examples

In the introduction to a 2017 Special Issue on "Doing Southern Theory," for *Postcolonial Directions in Education*, Takayama et al. provided the following examples of decolonial research methods being applied in practice using Southern Theory

> an innovative knowledge exchange project at the University of Western Sydney, Australia where Chinese and Indian higher degree research students' knowledge of intellectual and pedagogic work is fully utilized as a source of education theory development ... [and] a group of Asian education scholars at Monash University, Australia (Zhang, Chan & Kenway. 2015) [who] have drawn upon Kuan-Shin Chen's (2010) *Asia as method: Towards deimperialization* to develop alternative, Asia-focused methodological approaches to education research.

The model of faculty learning communities, academic fellowships, and other reflexive groups of scholars is a prime vehicle for exploring the role of AI in decolonizing CIE research. Similarly, Curtis (2023) used a decolonial approach called the Curtis Method to facilitate a study with Black feminist educators about how their spirituality influenced their teaching practices during the COVID-19 pandemic. The goal of this collective research was to develop a program and a paradigm to support Black women in postsecondary education by promoting liberatory pedagogical practices in the field and facilitating healing in the wake of the public health and social justice crises of the pandemic era. Despite the just aims of these studies and their intentionally decolonial guiding frameworks, these examples still center the academic knowledge production system and actors within it, namely faculty and graduate students. While these individuals are important, there are other qualitative methods of empirical research that CIE scholars may find helpful in structuring communities of inquiry about AI in research *outside* of the academic setting. The two examples listed below are just a few of the many frameworks for community-led research methods.

Participatory Research. Participatory research designs that "aim at positive, sustained transformations, as opposed to the stagnancy of data mining," offer CIE potential methodological alternatives to rely on (Nordtveit & Nordtveit, 2020, p. 34). Participatory research approaches can be applied to qualitative, quantitative, or mixed methods studies. As the name implies, in participatory research studies, "participants act to a great degree as co-researchers" (Merriam & Tisdell, 2016, p. 56). Participants are given the opportunity to not only determine the direction of the lead facilitators' study but also to act as action researchers themselves to conduct "studies in their own communities to specifically challenge power relations and initiate change in their own communities" (Merriam & Tisdell, 2016, p. 57). Participation in the development and execution of participatory studies "can affect and transform people from both an individual and societal perspective" (Merriam & Tisdell, 2016, p. 58).

Action Research. As a qualitative methodology, action research involves "partnering with and including" co-researchers in the study through the co-construction of research objectives, data collection processes, and validation of findings. According to Merriam and Tisdell (2016), there are four general types of action research: (i) technical action research, which aims to study ways to improve community control over research outcomes; (ii) practical action research, which aims to study ways to improve co-researchers behaviors and promote positive change from within a given community; and (iii) critical action research, which aims to study ways to work for community-led liberation from unjust circumstances. Lastly, participatory action research (PAR) gives community-members the opportunity to not only determine the direction of the study but also to act as researchers themselves. Unlike action research, which is generally "oriented toward some action or cycle of actions ... to address a particular problematic situation," in participatory action research, individuals have the full agency to conduct "studies in their own communities to specifically challenge power relations and initiate change in their own communities" (Merriam & Tisdell, 2016, p. 57). Furthermore, critical PAR studies "can affect and transform people from both an individual and societal perspective" (Merriam & Tisdell, 2016, p. 58).

Conclusion

Throughout this chapter, we have argued for the ways that AI can transform the field of CIE. However, we also concur with the OECD (2023), which states that "'while AI is penetrating all domains and stages of science, its full potential is far from realized.' The prize, it concluded, could be enormous: 'Accelerating the productivity of research could be the most economically and socially valuable of all the uses of artificial intelligence'" (Nolan, 2023, para 2). The primary barrier to AI's transformative potential may be sociological: that is, our progress with AI can happen only if we, as educators and international development specialists, are willing and able to use such tools. Therefore, we need to break the invisible researcher/ practitioner binary we have erected and use AI to truly study comparative systems, as AI changes the types of questions we are asking.

If CIE researchers are to use AI technology in ethical or socially just ways, let alone to further the project of decolonizing the field, we need to build our critical literacy. Critical literacy is the ability to "disrupt the commonplace, interrogating multiple viewpoints, focusing on sociopolitical issues, and promoting social justice" (Lewison, 2002, p. 382). CIE scholars who are committed to the project of decolonizing the field must learn how to examine and resist the reproduction of neocolonial practices in their research process and remain committed to critically examining and accounting for context in their work. We must also give more power to the community stakeholders that are served by research findings as the primary audience to whom we are accountable, ensuring their perspectives and desires are central in the research design and methods. At the very least, researchers using AI should obtain informed consent from participants before collecting and analyzing their data, and in their methodology, write-ups provide a thorough description of their process that explains the role of the AI tools in the research design and how harmful impacts of the tool(s) were mitigated. Unfortunately, these steps alone cannot account for the lack of influence that CIE researchers currently have in the development of these tools or the research behind them. Thus, more empirical studies with direct community involvement from the CIE lens are needed to examine the impact of AI tools on research practices in the field. Nonetheless, CIE research with and about AI in education also needs to interrogate the governance or regulatory frameworks emerging globally around AI and how they may or may not perpetuate colonial logics. This is a topic to which we turn next.

CHAPTER 6

Regulatory Responses and Emerging Global Scripts in the Governance of AI in Education

Abstract Looking at a selection of approaches to AI governance, this chapter provides an overview of the governance of AI in education (GAIE) and considers these through the lens of CIE research. We first begin with a brief look at education and the role AI might play in its execution. An examination of the selected AI frameworks from the European Commission, Türkiye, China, UNESCO, and the US Department of Education follows. Then, we consider some potential critical points of interest for CIE scholars as they relate to these and other frameworks before we provide some concluding thoughts about the role of CIE in continued interplay with developing AI capacities and deployment.

Keywords AI governance • AI in education • Regulation

Introduction

In the last decade, the development of machine learning (ML) models to enhance AI capacities has largely shifted from being primarily an activity of academia to one in which private industry has taken the lead (Maslej et al., 2023). This shift, which has been accompanied by seismic advances in AI technologies and capabilities and new consumer/private access to AI, has

© The Author(s), under exclusive license to Springer Nature
Switzerland AG 2024
S. M. S. Curtis et al., *The Technological-Industrial Complex and
Education*, https://doi.org/10.1007/978-3-031-60469-0_6

resulted in growing interest from policymakers to attend to both the perceived opportunities and threats posed by AI in many sectors, with a 650 percent increase in AI mentions in legislation globally since 2016 (Maslej et al., 2023). New AI developments and their current and forthcoming impacts on education might represent the latest in the ongoing string of changes to the "form, function, and mechanism" (Fan & Popkewitz, 2020, p. vi) of education.

This chapter is concerned with providing an overview of varied approaches to the governance and governing of AI in education (GAIE) based on a limited selection of frameworks for the governance of AI to date. It is essential to understand that one of the salient ideas in relation to the governance of AI in education is the extreme speed at which ML and AI technologies are developing. This has resulted, among other things, in policymaking falling quickly behind new AI capacities and approaches to policymaking that emphasize the importance of "future-proofing" legislative strategies so that policies developed today may be applicable to technologies and capabilities of the future (European Commission, 2021). In the context of AI, *future-proofing* has emerged as a way to mitigate the perceived risks and (to some extent benefits) that AI may pose. In introducing various AI governance frameworks from selected legislative and other bodies around the world, this chapter provides an overview of some of the approaches being taken to respond to AI opportunities and challenges. We first begin with a brief look at education and the role AI might play in its execution. An examination of the selected AI frameworks from the European Commission, Türkiye, China, UNESCO, and the US Department of Education follows this. We then provide consideration of some potential critical points of interest for CIE scholars as they relate to these and other frameworks before we provide some concluding thoughts about the role of CIE in continued interplay with developing AI capacities and deployment.

GOVERNING AI IN EDUCATION

This chapter focuses on GAIE as it relates to "formal" education, that is, learning institutionalized through formal schooling "in either primary education, or secondary education, or both and beyond" (Wiseman, 2022, p. 21). Engaging with the topic of the GAIE also necessitates an understanding of the various ends to which education is put; consideration of how these purposes may differ between countries, regions, and sub-populations; and examination of how AI in education may attend to or fail to attend to those diverse purposes. This question of focus is well outlined by Thomas (2013)

> Should schools primarily be about passing on knowledge and skills to a new generation, and if so which knowledge and which skills? Or should they put emphasis on the transmission of manners, habits, and traditions of culture? Should education be about encouraging compliance with the existing ideas and norms of a society or should it concern the promotion of questions, challenging, free-thinking disposition? (p. 16)

Additionally, should education be concerned with job training and labor productivity (OECD, 2023) or building global citizenship (Jules & Arnold, 2021)? Should education be geared towards the public or private good (Labaree, 1997; Williams, 2016), or something else entirely?

Local determination of the varied ends to which national and/or local education is put will necessarily be implicated in any party's approach to the GAIE, highlighting that consideration of local, regional, and national contexts is crucial for understanding how GAIE approaches may converge and diverge. As noted in previous chapters, the salience of local context obtains not only in the discussion of the purposes of education but also to the positioning of educational systems within global hierarchies, both real and imagined, as it relates to areas such as economic capacity, technology ubiquity, sovereignty and autonomy, human capital, power (in its many forms), and more.

AI in education can be further disambiguated through consideration of different areas within the topic. From a CIE perspective, one might identify four broad areas deserving of attention: (i) governance of AI in education (GAIE); (ii) AI in the classroom; (iii) AI for governing education; and (iv) AI for studying education. This chapter focuses specifically on the former, though the AI frameworks outlined below touch on topics associated with all four areas. When referring to governance, this chapter highlights some of the mechanisms used to outline expectations and norms of AI development, use, and deployment through the "management or leadership of the government" (Fasenfest, 2010, p. 771), where the "government" refers to bodies with varying levels of authority to enforce expectations and norms. This is a divergence from traditional forms of government, from "detailed regulation to framework legislation" (Hudson, 2007, p. 266), resulting in some cases in a decentralization that may incorporate more voices from both public and private entities (Hudson, 2007; Cole, 2008). Looking at the following AI frameworks from different countries and supranational bodies evidences a diversity of approaches to the challenges and opportunities posed by AI (in education). The varying approaches gesture not towards the abdication of leadership or government by legislative bodies but instead to the "state adapting to changing circumstances and finding new ways of governing" (Hudson, 2007, p. 266). This is especially salient due to the arguably dispersed nature of AI development across diverse sectors (public, private, and combined), countries, industries, and more.

FRAMEWORKS FOR AI GOVERNANCE

In this section, we cover four distinct frameworks for governing or strategizing vis-a-vis AI. This section identifies some of the broad strokes of AI governance and strategies apparent, providing insights into some of the ways that national, regional, and supranational bodies are taking up the task of responding to the challenges and opportunities posed by AI. As outlined in Table 6.1, the first two responses—from the European Commission and Türkiye's Digital Transformation Office—are AI policy documents not specific to education, instead considering the topic broadly, though touching on educational endeavors. These documents will impact national educational trajectories and define educational reforms around AI. The final three documents, from UNESCO, the People's Republic of China, and the United States Department of Education, focus squarely on AI in education.

The European Commission

One of the European Commission's most notable responses to AI is the 2021 proposal for an *Artificial Intelligence Act*, which was produced to "put forward legislation for a coordinated European approach on the human and ethical implications of AI" (p. 1). More specifically, the document provides a proposed framework to meet the following specific objectives

- ensure that AI systems placed on the Union market and used are safe and respect existing law on fundamental rights and Union values;
- ensure legal certainty to facilitate investment and innovation in AI;

Table 6.1 Selected AI governance frameworks

Title	Focus	Issuing body	Published
Proposal for a Regulation of the European Parliament and of the Council Laying Down Harmonised Rules on Artificial Intelligence (Artificial Intelligence Act) and Amending Certain Union Legislative Acts	Artificial intelligence	European Commission	2021
National Artificial Intelligence Strategy: 2021–2025	Artificial intelligence	Türkiye Digital Transformation Office	2021
Artificial Intelligence Innovation Action Plan for Institutions of Higher Education	Artificial intelligence and higher education	Ministry of Education, People's Republic of China	2018
Planning Education in the AI Era: Lead the Leap	Artificial intelligence and education	UNESCO	2019
Artificial Intelligence and the Future of Teaching and Learning	Artificial intelligence and education	Department of Education, United States of America	2023

- enhance governance and effective enforcement of existing law on fundamental rights and safety requirements applicable to AI systems;
- facilitate the development of a single market for lawful, safe, and trustworthy AI applications and prevent market fragmentation (European Commission, 2021, p. 3).

As a proposal for new AI regulations, the document works to define AI, install prohibitions on activities considered "harmful," and classify AI systems. Furthermore, it aims to create a "horizontal EU legislative instrument following a proportionate risk-based approach" to be combined with "codes of conduct for non-high-risk AI systems" (European Commission, 2021, p. 9). Mitigating risk, through prohibitions on particular AI practices (Title II) and through processes to identify high-risk uses of AI (Title III), is an essential feature of the proposal, and this is done with an eye on future-proofing ideas and concepts as AI technologies inevitably advance. The proposal also seeks to create "governance systems at [European] Union and national" levels (European Commission,

2021, p. 15), with Titles VI, VII, and VIII outlining the organization of these systems, a database for monitoring high-risk AI systems, and "monitoring and reporting obligations for providers of AI systems" for those deemed high-risk (European Commission, 2021, p. 15), respectively. The proposal seeks to balance the mitigation of risk with the insistence on an approach that will not unduly stymy AI innovation.

While focused on AI governance broadly, the proposal provides a selection of ideals as it relates to AI in education specifically. The Commission acknowledges the wide-ranging uses to which AI might be put to benefit society, including extending the benefits of predictive, optimizing, and personalizing capabilities to boost many areas, including "healthcare, farming, education and training..." (European Commission, 2021, p. 18), however, it cautions that one's training to use AI effectively is crucial for mitigating risks of AI, highlighting the need for education that prepares users for AI interactions across industries or fields. Finally, the proposal clarifies that unacceptable risks may accompany the use of AI in the administration of education and learning if proper safeguards are not put in place. Specifically, the proposal identifies as high-risk those educational activities that incorporate AI into decisions about "access or assigning persons to educational... institutions," as well as the use of AI to assess learning or to make placements (European Commission, 2021, p. 26). This raises a concern that AI, in these cases, may contain flaws that infringe on people's rights through the incorporation of discriminatory logic or algorithms (European Commission, 2021; Filgueiras, 2022).

Türkiye

The Turkish National Artificial Intelligence Strategy (NAIS) was published in 2021 by the Ministry of Industry and Technology and the President's Digital Transformation Office with the goal of "creating value on a global scale with an agile and sustainable AI ecosystem for a prosperous Türkiye" (Digital Transformation Office, 2021, p. 7). Core to the Strategy are the six diverse strategic priorities, accompanied by a defined "governance mechanism" for the "effective implementation of the NAIS" (Digital Transformation Office, 2021, p. 83). These strategic areas highlight the need for: (i) educating and employing more AI experts; (ii) backing research and innovation; (iii) enhancing data and technology capacities; (iv) developing regulations to augment adaptation to AI; (v) increasing international cooperation; and (vi) growing labor and structure changes (Digital Transformation Office, 2021). The document regularly refers to the development and growth of an ecosystem that will house the varied output components from the strategy. This ecosystem will comprise not only an amalgamation of the developed infrastructure, governance mechanisms, and expertise stemming from the strategy but it is also envisioned as the philosophical context within which this work will occur, founded on ideals of nimble "experimenting and implementing" (Digital Transformation Office, 2021, p. 8) new AI capacities and opportunities for these to transform Türkiye. The priorities outlined above point to the strategic, competitive orientation of the NAIS, as it seeks to capitalize on the opportunity afforded to those nations that quickly author and execute comprehensive approaches to engaging with and taking the lead on AI developments. Notably, the strategy highlights an approach to minimizing known and unknown risks of AI development through the implementation of outside regulatory approaches. Measure 1 of the fourth Strategic Priority ("Regulating to Accelerate Socioeconomic Adaptation") notes that "International regulations will be followed for the elimination of AI-related risks and the application of AI values and principles ..." (Digital Transformation Office, 2021, p. 72).

Contributing to the strategy's overall emphasis on a competitive approach to AI developments, the NAIS clarifies education as a central feature. Strategic priorities 1, 2, 3, and 6 all touch on the need to make changes to educational approaches and emphasize enhancing the country's capacity to engage in AI development, research, and application with an educated labor force. As with the European Commission's proposal, this strategy is not focused explicitly on the governance of AI in education, centering instead on the importance of education and associated research to benefit the country's AI approach (Digital Transformation Office, 2021). The strategy notes, for example, that "a radical transformation" (Digital Transformation Office, 2021, p. 45) is needed for education and training programs at the "pre-higher education" (p. 8) as well as post-secondary and post-graduate levels will be crucial. This is described as a project to harmonize education to meet the nation's AI needs, as the NAIS connects with other strategic educational initiatives already in place from other bodies. In short, mentions of education as it relates to AI within this strategy document are predominantly, or almost exclusively, geared towards capacity development and strategic competition in the field of AI, as opposed to the governance of AI in education.

The People's Republic of China

China's panoply of AI-focused or related policies covers a number of areas relevant to AI in education. From ethics to autonomous driving standards and from AI promotion activities to facial recognition protections, the country's AI plans and governance frameworks are truly eclectic and wide-ranging (OECD.ai). For example, in 2019 and 2021, the Ministry of Science and Technology published "governance principles" and "ethical norms," respectively, for the development and utilization of human-centered AI to outline ethical considerations (PRC Ministry of Science and Technology, 2021). These contribute to the growing body of Chinese policy and guidance documents pertaining to AI. Specific to the governance of AI in education, in 2018, the Ministry of Education issued an *Artificial Intelligence Innovation Action Plan for Institutions of Higher Education* (PRC Ministry of Education, 2018). This document prompts higher education institutions (HEIs) to "set their sights on the cutting edge of global science and technology (PRC Ministry of Education, 2018, p. 1)" and to pursue innovative AI development generally. The plan asserts the primal importance of HEIs as centers of innovative AI development that will support the "rejuvenation of China by embracing scientific education, strengthening China with talented people, innovation-driven development, and military-civil fusion" (PRC Ministry of Education, 2018, p. 3).

A core task for Chinese HEIs is the training and development of human capital to drive AI expertise in the nation. This will require institutions to conduct theoretical research, invest in software and hardware systems development, and pursue breakthroughs in a significant number of areas as they relate to core AI building blocks, such as "machine learning... deep reasoning, swarm intelligence... natural language intelligence" (PRC Ministry of Education, 2018, p. 5) and more. The plan furthermore implores HEIs to gather key talent that will drive this innovation, develop think tanks, and cultivate international relationships and collaborative efforts to enhance AI progress and development (PRC Ministry of Education, 2018).

The development of human and other resources to push AI boundaries and innovation will occur through improvements to training regimens through HEIs, including through new majors in the field efforts to augment teaching resources, among much else. Furthermore, breakthroughs and advancements in AI are envisioned as a stepping stone to incorporating AI into viable "commercial products" and championing the applicability of HEI research to other areas such as medicine, agriculture, "military-civil fusion," and other forms of cross-industry activities (PRC Ministry of Education, 2018, p. 10).

The policy emphasizes that to achieve these stated aims, it will rely on new bodies to "improve organization and implementation" (PRC Ministry of Education, 2018, p. 11) of the above efforts. Naturally, these HEI activities in AI innovation are expected to require significant resources, including, but not limited to, larger numbers of students to study AI and drive these envisioned innovations. The Ministry of Education is also tasked with creating new "innovation platforms," increasing funding, and endorsing "exploratory interdisciplinary research as well as foundational, forward-looking research to arrive at leading, original results and major breakthroughs" (PRC Ministry of Education, 2018, p. 13). Finally, the document outlines the Ministry's and local educational authorities' responsibilities to "increase propaganda on and promotion of major scientific and technological achievements to colleges and universities" (PRC Ministry of Education, 2018, p. 13).

UNESCO

The 2019 final report from UNESCO's International Conference on Artificial Intelligence in Education (*Planning Education in the AI Era: Lead the Leap*) contains a collection of broad recommendations for members, stakeholder organizations, and the UNESCO Director-General. In its calls for AI-specific policy development and specific uses for which AI should be put in education (UNESCO, 2019b), the report firmly centers "core principles of inclusion and equity" (Miao et al., 2021, p. 1). Naturally, for a body with the global scope of UNESCO, the final report highlights the possibility of "AI divides" mirroring current digital divides (UNESCO, 2019b), including in the realm of AI policymaking, as nations take different approaches or assign differing values to the idea of AI policy development Miao et al. (2021).

The report also seeks to frame the benefits and challenges posed by AI, identifying ways that it may be used as an "intelligence augmentation opportunity" for teachers and learners (UNESCO, 2019b, p. 20), though cautioning that there are limits to how AI should be used in education as it relates to less testable "competencies such as tolerance, intercultural understanding, and adaptability ..." (p. 21). Among much else, the report highlights the importance of matching technical progress with the development of the "social, legal and ethical dimensions of AI governance" as outlined within the document (UNESCO, 2021, p. 7); transparency will be critical in this regard.

In light of the challenges outlined as they relate to policymaking and governance of AI in education, the final report was succeeded by *AI and Education: Guidance for Policymakers* in 2021. This document is self-evidently oriented towards providing policymakers with ideas and practices to "leverage the opportunities and address the risks, presented by the growing connection between AI and education" (Miao et al., 2021). It overviews AI for policymakers, provides insights into the ways that AI converges with a number of education areas, and, crucially, provides policy recommendations. The Miao et al. (2021) guide's recommendations run the gamut but include strategic targets (e.g., "Ensuring the inclusive and equitable use of AI in education" [p. 31]), calls for a humanistic approach to AI in education, and identifies diverse stakeholders that may have a role to play in future developments and therefore should be included in guidance. The guide's emphasis on equity, ethics, and inclusivity touch on many areas of education, including "educational management, teaching, learning, and assessment" (p. 34).

The US Department of Education

Focused explicitly on the implications of AI on teaching and learning, the United States Department of Education's Office of Educational Technology produced a report titled *Artificial Intelligence and the Future of Teaching and Learning: Insights and Recommendations* (Cardona et al., 2023). This report explores the intricacies of dispatching AI in US classrooms through engagement with the challenges and benefits that may accompany such a move. As the title implies, the report provides a set of recommendations to guide policy development in this area.

Cardona et al. (2023) begin with an acknowledgment that interest in AI has grown considerably, that it has been accompanied by concerns about "issues of fairness and transparency," and that "research on ethics is increasing because problems are observed" (p. 3). The report works to respond to these and other concerns while also highlighting beneficial opportunities that might be pursued through AI-assisted teaching and learning. It also notes the importance of understanding how AI might interact or intersect with federal and state laws and is ultimately oriented towards informing the development of "ethical" and "equitable" policies (Cardona et al., 2023).

In pursuit of this aim, the report provides a number of four foundations upon which future policy on AI in education should rest. These foundations propose the importance of centering people, advancing equity, ensuring safety, ethical approaches, and effective implementation, and promotion of transparency (Cardona et al., 2023). These foundations are presented to contribute to the recommendations provided subsequent to the discussion of AI basics and AI as it connects to learning, teaching, assessment, and research.

The first recommendation holds one of the core tenets of the report, namely, the importance of solidifying the presence of humans within AI processes. This is presented with an apt analogy:

> We envision a technology-enhanced future more like an electric bike and less like robot vacuums. On an electric bike, the human is fully aware and fully in control, but their burden is less, and their effort is multiplied by a complementary technological enhancement. Robot vacuums do their job, freeing the human from involvement or oversight. (Cardona et al., 2023, p. 53)

The second and third recommendations seek to center human-determined educational goals and learning principles through the deployment of AI technologies that will facilitate achievement and adherence, respectively. This is outlined in an effort to ensure that AI remains a (powerful) tool that enhances the work of educators towards their specific aims for teaching and learning (Cardona et al., 2023).

The fourth and fifth policy development recommendations envision educational AI developments that "prioritize strengthening trust" (Cardona et al., 2023) while also centering the knowledge and involvement of educators, with both having the potential to dispel concerns that AI will simply replace and denigrate respect for educators. Finally, the sixth and seventh recommendations emphasize how crucial it is to approach AI with an eye on the specific needs of learners, educators, and the education endeavor as a whole, instead of pursuing the implementation into educational spaces of AI that is neither grounded in local contexts nor adhering to "education-specific guidelines and guardrails" (Cardona et al., 2023, p. 60).

CIE Scrutiny of AI Governance

The limited selection of AI frameworks provided above share with one another the balancing of optimistic exploration of opportunities afforded by AI with the importance of mitigating possible challenges and optimizing potential benefits from burgeoning capacities. They similarly share a core focus on guiding and controlling contemporary and future AI use and developments to meet the needs of identified stakeholders and to adhere to local or stakeholder expectations. Naturally, who these stakeholders are and what those expectations might be can vary depending on context, but these are shared, overarching concerns of these frameworks.

For example, the approaches from the European Commission, Türkiye, UNESCO, and the US Department of Education are concerned explicitly with the mitigation of risks that can accompany unchecked AI use and development. Türkiye and China, and perhaps less explicitly the European Commission, emphasize the importance of AI innovation for benefitting a number of market and industry areas. That is the capacity for deftly administered policies to not only guide AI use and development but also greatly enhance the AI competitiveness (broadly) of the countries and the EU as a region.

Each of the AI frameworks, in one way or another, concerns itself also with the development not only of AI infrastructure and innovation but also with the growth and advancement of human resources that are understood to be central to ensuring that AI use and development meets expectations for being human-centered, ethical, innovative, and competitive. UNESCO and the US Department of Education highlight the capacity of AI to enhance educational endeavors as a tool to augment that which is already occurring in the classroom, while they also identify the need for ethical foundations for any AI implementation projects.

The ongoing advancements of AI, as well as the new uses to which it is almost daily being put, prompt new questions and areas of inquiry for scholars of comparative and international education. When considering the governance of AI, scholars may consider paying special attention to the intersections of differing AI policy approaches within these and other frameworks. Broadly, one such area of inquiry may include seeking to understand what will occur at the intersections of (i) supranational/ regional governance approaches and (ii) national policy approaches? Additionally, AI is considered by many to be a disruptive collection of technologies in part because it can be fashioned for use in so many different areas, meaning that it is possible (or likely) that systems developed for one sector may be easily adjusted to work in others. For CIE scholars, this may prompt questions about the intersections of governance approaches to different AI emphases, such as industry, society, and governance of AI in education. It remains to be seen how well and quickly schemes for the governance of AI in education will be able to maneuver to respond to new opportunities and challenges that arise from new AI development and use areas that are not specifically designed for education.

The potential for continued and ubiquitous AI disruptions across many areas, including education, should prompt CIE scholars to consider how new and developing AI systems, technologies, and potential—and subsequent policy and governance responses—might affect their specific areas of interest within the field. This is an acknowledgement that AI technologies are being explored in a vast number of areas and that this is sure to begin affecting (if it has not already) almost all areas of education, though perhaps in uneven and inequitable ways across the globe.

An example is useful here to illustrate how scholars might approach AI impacts on their areas of research. Some CIE researchers focus on international mobility and immigration policies pertaining to flows of students and scholars across national borders. As such, these scholars might explore the impacts of competitive governance approaches on the development of AI in the United States on international mobility and immigration policy.[1] Brunner and Tao (2024) discuss how AI is "newly impacting Canada's governance of international students" (p. 271), while in an earlier study, Grimm (2019) undertook a comparative study exploring how state policy development is used to "enable or restrict international student graduates from domestic institutions to be able to stay and work beyond their studies or transition towards permanent status..." (p. 236). A later comparative article then showed how immigration policies for students can influence their decisions about what to study, with, for example, visa policies incentivizing some students to study science, technology, engineering, and mathematics (STEM) in the United States when they might otherwise have focused in different areas (Crumley-Effinger, 2024).

Mirroring some of the discourse in the governance frameworks from Türkiye and China, a late 2023 executive order (Executive Order on the Safe, Secure, and Trustworthy Development and Use of Artificial Intelligence) from President Joseph Biden indicates how federal

---

[1] Scholars looking at these topics may, for example, be part of CIES's Global Migration (SIG #13) and/or Study Abroad and International Students (#27) special interest groups. See Table 6.2 below.

immigration policy can be used as one approach to contending with new imperatives in the global competition for AI development leadership. Section 5 of the executive order describes how immigration policies affecting students, scholars, and researchers must be updated to, among other things,

> streamline processing times of visa petitions and applications, including by ensuring timely availability of visa appointments, for noncitizens who seek to travel to the United States to work on, study, or conduct research in AI or other critical and emerging technologies .... (Exec. Order No. 14110, 2023)

These outlined efforts are accompanied by an expectation that new immigration-related changes should be made that will benefit certain populations of scholars and students who are studying in STEM fields, which is done, naturally, with the expressly stated purpose of "attracting AI talent to the United States" (Exec. Order No. 14110, 2023). As noted previously, China will also seek augmentation of its AI innovation and development through increased "support for exchange students in the AI field" (PRC Ministry of Education, 2018, p. 8), while Türkiye will similarly pursue "mutual student, expert and academician exchanges with other countries" (Digital Transformation Office, 2021, p. 76).

While the immigration implications of these two exchange emphases were not explored in the provided frameworks from their respective countries, they point to the ways in which CIE scholars focusing on international student and scholar mobility will contend with the broad impacts of governance of AI in education in varying ways. The above US executive order and other examples show how national AI governance and policy approaches might impact one small corner of the field of CIE (international mobility and immigration policy), exemplifying what will likely be an area of increasing importance for AI and CIE scholarly inquiry in the coming years.

The CIE focus on international mobility is just one example of the ways that CIE scholars may be confronted with a new prerogative to conduct research and develop theory to explore how AI developments, use, and governance impact education globally. Guided by the special interest groups (SIGs) of the Comparative and International Education Society (CIES), it is fruitful to consider other examples of potential new or ongoing CIE research as it relates to the governance of AI in education and its current and ongoing aftereffects (see Table 6.2). What follows are a few examples of groupings of CIES SIGs that may take particular interest in

Table 6.2   CIES topic-based SIGs (from CIES, n.d.)

#	SIG name	#	SIG name
1	African Diaspora	16	ICT4D [ICT for Development]
2	Citizenship and Democratic Education	17	Inclusive Education
3	Contemplative Inquiry and Holistic Education	18	Indigenous Knowledge and the Academy
4	Cultural Contexts of Education and Human Potential	19	Language Issues
5	Early Childhood Development	20	Large-Scale Cross-National Studies in Education
6	Economics and Finance of Education	21	Monitoring and Evaluation
7	Education, Conflict, and Emergencies	22	Peace Education
8	Educational Improvement	23	Philanthropy and Education
9	Environmental and Sustainability Education	24	Post-foundational Approaches to CIE
10	Gender Justice	25	Religion, Education, and Society
11	Global Literacy	26	Sexual Orientation and Gender Identity and Expression
12	Global Mathematics Education	27	Study Abroad and International Students
13	Global Migration	28	Teacher Education and the Teaching Profession
14	Globalization and Education	29	Teaching Comparative and International Education
15	Higher Education	30	Youth Development and Education

engaging with some of the themes arising in the provided governance frameworks. The governance frameworks from both China and Türkiye emphasize new funding exigencies as they relate to education to develop AI human talent, research, and innovation. These new funding plans and expectations will likely be interesting for CIE scholars in many different areas, but perhaps particularly with implications for those in the following SIGs: Economics and Finance of Education (#6), Global Mathematics Education (#12), and ICT4D (#16). The frameworks issued by UNESCO and the US Department of Education focus considerable energies on the importance of centering ethics, equity, and inclusion in the dispersal of AI in the classroom, which will likely pique the interest of scholars focusing on topics coming under the umbrellas of Educational Improvement (#8), Inclusive Education (SIG #17), and Teacher Education and the Teaching Profession (#28), among others.

These are just a selection of examples of how contemporary and future AI governance frameworks might prompt new inquiry by CIE scholars across many issue areas within the field. Ideally, the examples provided here might spark consideration for those with specific areas of expertise, prompting them to think of many more relevant considerations as they relate to the ideas presented in the example frameworks and in previous chapters. This book has emphasized the importance of equity, decoloniza-tion, and epistemic inequity vis-a-vis AI's background, development, and use. One might, therefore, consider how scholars with a particular interest in, for example, Indigenous Knowledge and the Academy (SIG #18), Gender Justice (#10), and Citizenship and Democratic Education (#2) may be particularly keen to explore the impacts of new AI developments and governance frameworks in these different areas.

CONCLUSIONS

AI presents education and those studying it with a vast array of new oppor-tunities, challenges, and potential. As is evidenced by some of the frame-works provided here—as well as contemporary discourse on the opportunities and threats posed by AI—reactions to new AI development run the gamut of approaches and views on the future of this large umbrella of technological capacities. That frameworks such as these are trying to develop and then put in place structures to attend to a prodigiously fast-changing collection of AI systems often developed in the private sector shows how challenging the task is. An emphasis on future-proofing gover-nance frameworks also highlights the concerns with keeping governance in line with new developments, and the potential for valuable and valued opportunities afforded by these new systems also brings to bear the urgency with which national and supranational bodies are attending to these issues.

Frameworks such as these are coming to the fore in the face of oppor-tunities, swift technological change, increased concern about data privacy and ownership, and as different entities seek to respond to the genuine differences in ICT capacities impacting AI access. These are often consid-erations of *context*. A particular nation's or region's context as it pertains to the development and implementation of AI systems necessarily shapes their approaches, and CIE scholars may be uniquely positioned to support ongoing efforts to come to terms with the challenges and opportunities posed by AI systems that have been developed in certain contexts, and that may be implemented in others. This, again, brings the following questions of context to the fore, in this case, specifically about the US of AI in education matters:

- Who is developing the AI systems?
- For whom are the systems being developed?
- To whom will the systems ultimately be deployed and how will the local context be incorporated in this deployment?
- From what data will the systems be trained and developed?
- How will student and educator data be used, both in terms of data privacy and in terms of informing future iterations of systems?

This is by no means a comprehensive list of questions that may be of particular interest to CIE scholars, but these questions may be particularly relevant to CIE scholars concerned with the governance of systems and technologies used in education. It will not do to mirror biased AI technology development with governance scripts that continue to impart biased and region- or nation-specific conceptions of education through single-minded or one-size-fits-all approaches to outlining the contours of AI in education.

CIE may also be particularly well-positioned to contend with complicated questions at the intersections of GAIE and equity, social justice, and decolonization. Scholars in the field have begun to contend with the international dimensions of many aspects of injustice that have flowed from the educational, political, and intra-state systems that have made education what it is, in all its diversity, around the world. This includes concerns about educational access, quality, equity, borrowing, assessment, philosophical geneses, and more. In short, while CIE scholars will likely have little to no role in developing AI systems at the organizational, technological, or private enterprise level, it may become increasingly important for CIE researchers to use their voices and scholarship to support, critique, and envision governance frameworks working towards an ethical, equity-focused, and decolonial AI-influenced educational future.

CHAPTER 7

Capturing the Potential of Pluriversal AI Ecosystems

Abstract In this final chapter, we conclude by offering a summary of the different chapters. This is followed by a discussion on decoloniality and Pluriversal thinking. We return to the idea of what it means to decolonize AI in the era of the Fourth Industrial Revolution and its implications for the field of CIE. We conclude by offering some thoughts for the future of AI in CIE.

Keywords Epistemologies • Cosmovisions • Pluriversal

It is difficult to conclude this book in that we do not have a crystal ball to predict the future of AI and how it can/will genuinely shape the field of CIE in the anthropogenic era of the Fourth Industrial Revolution. What we are calling for is a more human-centered approach to the advancement, development, and deployment of AI in education. Such an approach would respect Indigenous knowledges and cosmovisions and comprehend that the current data that AI is trained upon is inherently biased, and this bias must therefore be taken into consideration in the deployment of AI in education, in the use of AI in CIE research, and more. In the introductory chapter, we raised a set of thought-provoking questions about the implications of AI for social justice in education, setting the stage for further exploration of these issues. Throughout the book, we went on to address

S. M. S. Curtis et al., *The Technological-Industrial Complex and Education*, https://doi.org/10.1007/978-3-031-60469-0_7

these questions and explained their potential impact on CIE. This book focuses on the ways in which CIE scholars can work to disrupt colonial practices embedded in AI. In this chapter, we offer a summary of the previous chapters and some preliminary thoughts on the future of AI and its efficacy on CIE. We also return to the idea of decolonization, particularly the notion of pluriversality, and discuss its potential implications for AI. We conclude with some ideas about the future of AI in CIE.

Chapter 2 talked about AIED and CIE from a social justice lens and explored the development and impact of AI on CIE. We discussed AI's role in mimicking human intelligence and explored the rise of the technological-industrial complex and the transition from Education 1.0 to Education 4.0. Our discussion emphasized the need to decolonize AI, and address existing structural inequalities to prevent the reinforcement of historical inequalities in AI training data. It also highlights the impact of the COVID-19 pandemic on the rapid advancement of AI in education and the shift towards a digital society.

Chapter 3 examined the historical and cultural roots of AI, from ancient mythology to modern science fiction, and its potential applications in education. It covered the portrayal of AI in popular culture and its real-life development, including the skepticism and challenges surrounding AI. The influence of AI on CIE and its potential implications for social justice are also explored. The concept of artificial neural networks and the different types of AI are discussed, along with their potential impact on education. Ethical concerns about the use of AI in education, including commercialization and cultural bias, are highlighted, emphasizing the need for critical examination and regulation of AI tools in education to safeguard the humanity and dignity of educators and learners.

Chapter 4 considered the ethical considerations of using AI in CIE, emphasizing the need for ethical guidelines for responsible AI use. It addressed concerns about digital coloniality, equity, and access. We highlighted the Manifesto on Digital Humanism and advocated for the design of technologies in line with human needs and values, emphasizing democracy, inclusion, and human decision-making in AI development. The advantages of AI in educational research and practices are outlined, such as access to diverse data and breaking down language barriers, but ethical implications of unequal access and perpetuating educational inequalities are noted. The challenges of incorporating AI in CIE, including the potential marginalization of alternative knowledge and ethical considerations of implementing AI, are detailed.

Chapter 5 emphasized the need for scholars to apply ethical and community-centered practices when using AI in research. It discussed the potential for AI to transform the field of CIE and highlighted the importance of critical literacy, community involvement, and obtaining informed consent when using AI tools. The chapter also addressed the methodological implications of AI in CIE research, including the impact on research methods, ethical considerations, and differences in epistemologies between CIE and AI. Additionally, it explored the implications of AI for decolonizing CIE research and emphasized the limitations of AI in replacing the perspectives and will of CIE scholars. The chapter concluded with examples of participatory research and action research as alternative methods for community-led research and provides references to support its arguments.

Chapter 6 provided an overview of the governance of AI in education, discussing the shift of AI development from academia to private industry and the growing interest from policymakers in addressing the opportunities and threats posed by AI in education. It examined the strategic priorities and mechanisms for governing AI across different countries and supranational bodies. Additionally, it highlighted the potential impact of AI governance frameworks in CIE, emphasizing the importance of considering equity, decolonization, and epistemic inequity in the research on and implementation of AI systems in education.

BEYOND EXTRACTIVISM: CLOSING AND OPENING

Throughout this book, we have been guided by the question of *how CIE scholars and practitioners can use AI to decolonize research and practice in CIE*. In an effort to answer this question, we have examined the quick pace at which AI is advancing and its implications for CIE. We have argued that CIE is a colonial project that is in the midst of decolonizing, and as such, it needs to pay attention to the myriad ways in which AI can be used in the field. AI is here to stay, and it is not just another passing fad, nor is it a panacea for the current inequalities that plague national educational systems. This warrants that we pay attention to how AI can be used ethically and socially to achieve inclusion and equity.

Our primary concern with AI today is that as algorithms have grown considerably more complex, the data that is used to train AI still has problems. Bias can creep into algorithms in several ways and cause harm as AI systems do not operate in isolation. A complete understating of biases should account for human, systemic, and computational biases. While human bias comes about through the ways in which people use data, systemic biases occur when institutions operate in ways that disadvantage certain social groups, such as discriminating against individuals based on their race, gender, or sexuality. We understand that current data sets are flawed in that they are inherited from historical, social inequities, and human bias . One way to rectify this is by ensuring that AI is trained on data that recognizes historical flaws of Western-centric notions of Cartesian ideas of progress and modernity. While AI cannot be genuinely bias-free, we can work to minimize algorithmic data bias by paying attention to the learning models that are written by people and socially generated data that is used to train AI algorithms. In looking for data bias in CIE, we may want to consider that data gathering should be structured in such a way that it allows for the inclusion of different knowledge options. This includes questioning what types of knowledges are used in the data set, where this knowledge is generated from, and what ethical considerations must be borne in mind if this data is to be used.

It is important to remember that AI is not designed to replace humans; however, it is already changing the ways we do research and development work by bolstering innovation, productivity, and outcomes. As such, we need to understand the possibilities and risks that AI brings while preparing for the changes that it will bring about in education. We need to ensure that we are creating a future where AI serves as a tool for the advancement of research and practice in the field of CIE. AI will require that researchers and practitioners do a skill shift as part of their future-proofing strategies to navigate the challenges and opportunities posed by AI. We, therefore, need to address the concerns with and resistance to AI openly while anticipating the changes that are coming. The integration of AI has the potential to redefine research and development work.

Throughout this book, we have discussed the different changes that AI has brought about and the ways in which they can influence the field of CIE. After exploring the historical, cultural, philosophical, and ethical dimensions of AI in education, we call for a critical examination and regulation of AI tools in education to safeguard the humanity and dignity of educators and learners, particularly in the context of CIE. We have outlined the potential implications and challenges associated with the integration of AI in CIE. AI has transformative potential for CIE, and as such, there is a need to prioritize research and learning outcomes over digital inputs and to ensure that AI is used as a complement to face-to-face strategies in education. Despite the prevailing challenges, AI can be beneficial to

CIE scholars by enhancing research productivity and transforming the potential applications of their findings. In using AI technologies in CIE, researchers should strive to ensure that diversity and accessibility are an integral part of our projects.

In this book, we touched on two crises (the ecological crisis and the AI crisis) that threaten to lead humans down a perilous path. Given that AI is created within the context of global power hierarchies and the privileging of Eurocentric innovations, pedagogies, and instruction and functions in a neoliberal market atmosphere, we argue that we need to begin with unpacking, un-anchoring, and understanding the colonial legacies of modern Western imperialism to move beyond the "epistemic chokehold of the [Global] North" in CIE. (Sicka & Hou, 2023, p. 28). In other words, the effects of colonialism are profound and continuing, and we have to find ways to undo some of the damage even if we cannot undo the past experiences. Therefore, we need to move away from "reading from the center" (Connell, 2007, p. 44) and beyond what Byrd (2013) calls "colonial agnosia," the uneasiness of unlearning and unknowing, given that neoliberalism has become discursively constituted to meet the requirements of modernization in the context of Euro-modernity. The decolonization of AI represents the changing geopolitics of knowledge where the modern onto-epistemological framework of knowledge is no longer tied to Western notions of modernity and progress and, as such, we should account for human, systemic, and computational biases. In addition to algorithmic bias, using AI in CIE research also raised ethical concerns around data privacy and the increasing digital divide in that current forms of AI do not account for broader sociopolitical and historical forces that recast transnational inequalities.

The depth of the contemporary crises (ecological, capitalist, epistemic, etc.) that we are facing warrants new ways of dealing with them. The future of intelligence demands that we properly harness the transformative potential of AI as we continue to rethink how to study educational systems from a comparative perspective. Such a way of thinking moves beyond globalization, and it "transcends the universalization of modernity into an understanding of planetarization as the creation of better conditions for the pluriverse" (Escobar, 2018, p. 64). An ontological reading of the social movements shows that such transiting is occurring in the Global South, and AI developers may want to consider how these transformations can shape its development. As we think about how to move from the "one-world Euro-American metaphysic" (Law, 2011) and modern dualist to pluriversal thinking, we must consider the "geoepistemopolitics" of what Escobar (2018) calls the transition discourse. From a relational onto-logical perspective, AI offers CIE researchers numerous possible worlds that can carry the field in multiple directions. However, these directions are all interconnected and overlap with each other. Nonetheless, we need to be careful of the pitfalls that AI can bring to research in CIE. Given the potential catastrophe(s) of the Anthropocene, CIE researchers should conceive of a world of many worlds where they must understand the ways in which the present world systems are conceived of and driven by neolib-eral market fundamentalism and economic globalization. This implies that we cannot see research as only extractive and may think of how it can contribute to engendering different types of developments that are not capitalist.

With the rise of large-scale datasets in CIE, educational research has become multidimensional. In the coming years, our reliance on these large-scale datasets will only increase, and we will need to pay attention to the ways in which data is collected for these datasets. New datasets will be generated from new data sources and will be more sophisticated and com-plex, and as such, traditional data processing software will not be able to manage them. From a research perspective, data innovation will allow us to study the linkages between humans, institutions, entities, and processes and then determine new ways to use those insights. As researchers focus on how research-based knowledge is conceptualized and produced, they need to remember that this will involve developing "alternative methodol-ogy as non-totalizable, sometimes fugitive, also aggregate, innumerable, resisting stasis and capture, hierarchy and totality, what Deleuze might call 'a thousand tiny methodologies'" (Lather, 2013, p. 635).

As we have argued, AIED is not new, and it holds promise for the field of CIE. Since November 2022, with the launch of ChatGPT, renewed attention has been placed on AIED. ChatGPT represents disruptive inno-vation in the educational landscape, but it also serves as a reminder that AI has been functioning in education for some time through conversational robots and natural language processing, learning analytics, speech and visual recognition, adaptive learning, decision support, and systems expert systems (Poellhuber et al., 2024). Before ChatGPT, not many scholars thought about AIED in CIE. Much of the earlier work on AIED was on adaptive learning, learning analytics, and the use of LMS such as Moodle, Blackboard, Sakai, Canvas, Google Classroom, and Schoology. However, these platforms were not used to conduct research *per se*, and the focus was on how the incorporation of educational resources (files, HTML pages, hyperlinks, etc.) and the use of assessments aided the learner, improved learning, and fostered a conducive learning environment. Work in this area fell into four categories: "performance prediction, decision support for teachers and learners, behavioral pattern detection and learner modeling, and dropout prediction" (Poellhuber et al., 2024, p. 153), and issues around personal data and confidentiality arose as challenges. The arrival of ChatGPT, which can create content (texts, computer code, images, videos) from prompts and language processing, heralded a new dimension of generative AI in education, which allows researchers to synthesize existing data or summarize ideas or present data differently and it can also function as an adaptive research assistant. However, AIED has limits, and the technology is still evolving. The emergence of deep fakes that use AI to manipulate real people's voices and/or videos in fictitious situations or discussions is an issue which needs to be tackled. Other risks the field needs to be aware of when using AI include ethical manipulation, plagiarism, data privacy, biases, lack of transparency, and data dependency. However, the benefits of AI may outweigh its detriments as it has the potential to transform the development of research skills while helping us to solve some of the Anthropocene's most wicked problems. In the meantime, we need to advocate for more human-centered AI, and measures must be taken to ensure research transparency in the field of CIE. Only when all these factors are considered will we see ethical, relevant, and long-term changes in CIE research.

The existential risks that AI pose have fascinated and frightened people. The canonical example often cited is the destruction of humans. The application of AI to CIE has several consequences that we have highlighted across this book. Several scholars have argued that computers will become better and move away from being under human control. These advanced systems will employ "tool use" as computers learn to do things better. With so many open-source unregulated technologies on the market today, it is hard to predict how different AI tools will need to be constrained so that it does not do harm. We must remember that the harm that AI can cause is a product of human training and therefore to mitigate harm, AIs need to be properly trained and be more human-centered. We advance that we need to develop a research monitoring group for the field that creates an "educational software architecture" that spells out the ethical use of AI in education and, by extension, CIE. Such a step will help us to think about the role of AI in educational research. We need to build educational AI with ethical principles that must always be followed. AI has come a long way, and it has a long way to go in education. As we think about decolonizing the field of CIE in an era of deep fakes, chatbots, and post-truths, we may want to establish guardrails to protect research subjects and researchers alike. While we have focused on AI, the cometh of artificial general intelligence (AGI), which is the ability of AI to think like humans, be self-aware, and be conscious, raises concerns for educational research. AGI has already found usage in education ranging from creating dynamic learning environments to providing feedback to improving computational thinking skills. Like AI, AGI too will need to be properly deployed, managed, and regulated if it is to benefit humanity and use education to solve some of the globe's wicked problems. But for now, AI can move CIE out of its positivist-interpretivist split and usher in a golden age of research discovery for the field that fosters human-centered models that are guided by the principles of ethics, justice, and inclusion. Thus, as we challenge the limits of the social continuum, we have to be careful that we do not succumb to what we seek to confront. Therefore, as we move away from the privileging of subjectivity and towards placing the human on an immanent plane, in so doing ridding it of its ontological privilege, or what has been termed new materialism, we must recognize such a move holds tremendous promise for the use of AI in the field of CIE.

REFERENCES

Adami, C. (2021). A brief history of artificial intelligence research. *Artificial Life*, *27*(2), 131–137. https://doi.org/10.1162/artl_a_00349

Adams, R. (2021). Can artificial intelligence be decolonized? *Interdisciplinary Science Reviews*, *46*(1–2), 176–197. https://doi.org/10.1080/0308018 8.2020.1840225

Alagraa, B. (2018). Homo Narrans and the science of the word: Toward a Caribbean radical imagination. *Critical Ethnic Studies*, *4*(2), 164–181. https://doi.org/10.5749/jcritethnstud.4.2.0164

Asimov, I. (1950). *Runaround: I, Robot* (The Isaac Asimov Collection ed.). Doubleday.

Berners-Lee, T. (2017). Three challenges for the web, according to its inventor. https://webfoundation.org/2017/03/web-turns-28-letter/

Birhane, A. (2019). The algorithmic colonization of Africa. *Real Life Magazine*. https://reallifemag.com/the-algorithmic-colonization-of-Africa/

Bon, A., Dittoh, F., Baart, L. A., Pini, M., Bwana, R., WaiShiang, C., & Kulathuramaiyer, N. (2022). Decolonizing technology and society: A perspective form the Global South. In H. Werthner, E. Prem, E. A. Lee, & C. Ghezzi (Eds.), *Perspectives on digital humanism* (pp. 61–68). Springer.

Bowers, C. A. (2014). *The false promises of the digital revolution: How computers transform education, work, and international development in ways that undermine an ecologically sustainable future*. Peter Lang.

Bray, M., Adamson, B., & Mason, M. (2016). *Comparative education research: Approaches and methods*. Springer.

Brickman, W. W. (1950). International education. In W. S. Monroe (Ed.), *Encyclopedia of educational research*. Macmillan.

S. M. S. Curtis et al., *The Technological-Industrial Complex and Education*, https://doi.org/10.1007/978-3-031-60469-0

Brickman, W. W. (1966). Prehistory of comparative education to the end of the eighteenth century. *Comparative Education Review, 10*(1), 30–47.

Brown, J. S., Collins, A., & Duguid, P. (1989). Situated cognition and the culture of learning. *Educational Researcher, 18*(1), 32–42. https://doi.org/10.3102/0013189X018001032

Brunner, L. R., & Tao, W. W. (2024). Artificial intelligence and automation in the migration governance of international students: An accidental ethnography. *Journal of International Students, 14*(1), 269–288. https://doi.org/10.32674/jis.v14i4.5762

Bu, L. (1997). International activism and comparative education: Pioneering efforts of the International Institute of Teachers College, Columbia University. *Comparative Education Review, 41*(4), 413–434.

Buchanan, B. G. (2005). A (very) brief history of artificial intelligence. *AI Magazine, 26*(4), 53–60. https://doi.org/10.1609/aimag.v26i4.1848

Bunnell, T. (2006). The growing momentum and legitimacy behind an alliance for international education. *Journal of Research in International Education, 5*(2), 155–176.

Byrd, J. (2013, October 23). *Silence will fall: The cultural politics of colonial agnosia* [research seminar]. WGSI Research Seminar, University of Toronto.

Cambridge, J., & Thompson, J. J. (2004). Internationalism and globalization as contexts for international education. *Compare, 34*(2), 161–175.

Campbell, M., Hoane, A. J., Jr., & Hsu, F. H. (2002). Deep Blue. *Artificial Intelligence, 134*(1–2), 57–83. https://doi.org/10.1016/S0004-3702(01)00129-1

Cardona, M. A., Rodríguez, R. J., & Ishmael, K. (2023). *Artificial intelligence and the future of teaching and learning: Insights and recommendations.*

Carter, S. M., & Little, M. (2007). Justifying knowledge, justifying method, taking action: Epistemologies, methodologies, and methods in qualitative research. *Qualitative Health Research, 17*(10), 1316–1328.

Césaire, A. (2001). *Discourse on colonialism* (J. Pinkham, Trans.). Monthly Review Press.

Chassignol, M., Khoroshavin, A., Klimova, A., & Bilyatdinova, A. (2018). Artificial intelligence trends in education: A narrative overview. *Procedia Computer Science, 136*, 16–24.

Chen, L., Chen, P., & Lin, Z. (2020). Artificial intelligence in education: A review. *IEEE Access, 8*, 75264–75278. https://doi.org/10.1109/ACCESS.2020.2988510

Christensen, C. (1997). *The innovator's dilemma: When new technologies cause great firms to fail.* Harvard Business School Press.

Christensen, C. M., McDonald, R., Altman, E. J., & Palmer, J. (2016). Disruptive innovation: Intellectual history and future paths. Working Paper 17–057. Harvard Business School.

Christensen, C. M., & Raynor, M. (2003). *The innovator's solution: Creating and sustaining successful growth.* Harvard Business School Press.

CIES. (n.d.). Special Interest Groups. https://cies.us/special-interest-groups/

Clemens, M. A. (2008). The long walk to school: International education goals in historical perspective. *SSRN Electronic Journal, 37.* https://doi.org/10.2139/ssrn.1112670

Close, K., Warr, M., & Mishra, P. (2023). The ethical consequences, contestations, and possibilities of designs in educational systems. *TechTrends.* https://doi.org/10.1007/s11528-023-00900-7

Cole, A. (2008). *Governing and governance in France* (No. 16829). Cambridge University Press.

Connell, R. (2007). *Southern theory.* Polity Press.

Consensus. (2024). *AI search engine for research.* Consensus. https://consensus.app/.

Coppin, B. (2004). *Artificial intelligence illuminated.* Jones & Bartlett Learning.

Cortina, R. (2019). Presidential address: "The passion for what is possible" in comparative and international education. *Comparative Education Review, 63*(4), 463–479.

Cowen, R. (2000). Comparing futures or comparing pasts? *Comparative Education, 36*(3), 333–342.

Cram, F. (2015). Harnessing global social justice and social change with multiple and mixed method research. In S. N. Hesse-Biber, & R. B. Johnson (Eds.), *The Oxford handbook of mixed and multiple methods research* (pp. 667–687). Oxford University Press.

Crawford, K. (2021). *Atlas of AI: Power, politics, and the planetary costs of artificial intelligence.* Yale University Press.

Creutzig, F., Acemoglu, D., Bai, X., Edwards, P. N., Hintz, M. J., Kaack, L. H., Kilkis, S., Kunkel, S., Luers, A., Milojevic-Dupont, N., Rejeski, D., Renn, J., Rolnick, D., Rosol, C., Russ, D., Turnbull, T., Verdolini, E., Wagner, F., & Wilson, C. (2022). Digitalization and the anthropocene. *Annual Review of Environment and Resources, 47*(1), 479–509. https://doi.org/10.1146/annurev-environ-120920-100056

Crumley-Effinger, M. (2024). ISM policy pervasion: Visas, study permits, and the international student experience. *Journal of International Students, 14*(1), 78–96. https://doi.org/10.32674/jis.v14i1.5347

Crutzen, P. J. (2002). Geology of mankind. *Nature, 415*(6867), 23.

Curtis, S. (2023). *Pandemic pedagogies: Fatalistic or black feminist?* ProQuest.

de Winter, J. C. (2023). Can ChatGPT pass high school exams on English Language comprehension? *International Journal of Artificial Intelligence in Education.* https://doi.org/10.1007/s40593-023-00372-z

Digital Transformation Office. (2021). National artificial intelligence strategy 2021–2025. https://cbddo.gov.tr/en/nais

Dobrin, S. (2023). Talking about Generative AI: A guide for educators. https://sites.broadviewpress.com/ai/talking/.

Dolby, N., & Rahman, A. (2008). Research in international education. *Review of Educational Research, 78*(3), 676–726.

Doroudi, S. (2022). The intertwined histories of artificial intelligence and education. *International Journal of Artificial Intelligence in Education.* https://doi.org/10.1007/s40593-022-00313-2

Drucker, P. (1969). *The age of discontinuity: Guidelines to our changing society.* Harper and Row.

Du Boulay, B. (2023). Artificial intelligence in education and ethics. In O. Zawacki-Richter & I. Jung (Eds.), *Handbook of open, distance and digital education* (pp. 93–108). Springer.

Epstein, E. H. (2008). Setting the normative boundaries: Crucial epistemological benchmarks in comparative education. *Comparative Education, 44*(4), 373–386. https://doi.org/10.1080/03050060802481405

Erscoi, L., Kleinherenbrink, A. V., & Guest, O. (2023). Pygmalion displacement: When humanising AI dehumanises women. https://doi.org/10.31235/osf.io/jqxb6

Escobar, A. (2018). *Designs for the pluriverse: Radical interdependence, autonomy, and the making of worlds.* Duke University Press.

Escobar, A. (2020). *Pluriversal politics: The real and the possible.* Duke University Press.

EU High-Level Expert Group on AI. (2019). Ethics guidelines for trustworthy AI. https://www.ccdcoe.org/uploads/2019/06/EC-190408-AI-HLEG-Guidelines.pdf

European Commission. (2019). Ethics guidelines for trustworthy AI, Publications Office. https://data.europa.eu/doi/10.2759/346720

European Commission. (2021). Proposal for a regulation of the European Parliament and of the council laying down harmonised rules on artificial intelligence (Artificial Intelligence Act) and Amending Certain Union Legislative Acts. https://eur-lex.europa.eu/legal-content/EN/TXT/?uri=celex%3A52021PC0206

Exec. Order No. 14110, 88 FR 75191. (2023). https://www.whitehouse.gov/briefing-room/presidential-actions/2023/10/30/executive-order-on-the-safe-secure-and-trustworthy-development-and-use-of-artificial-intelligence/

Fan, G., & Popkewitz, T. (2023). Introduction: Education policy and reform in the changing world. In G. Fan & T. S. Popkewitz (Eds.), *Handbook of education policy studies* (pp. v–xx). Springer Nature.

Fasenfest, D. (2010). Government, governing, and governance. *Critical Sociology, 36*(6), 771–774.

Filgueiras, F. (2022). Artificial intelligence policy regimes: Comparing politics and policy to national strategies for artificial intelligence. *Global Perspectives, 3*(1), 32362.

Filgueiras, F. (2023). Artificial intelligence and education governance. *Education, Citizenship and Social Justice* (Online). https://doi.org/10.1177/17461979231160674

Fishel, S. R. (2017). *The microbial state: Global thriving and the body politic.* University of Minnesota Press.

Fisk, P. (2017). Education 4.0 … the future of learning will be dramatically different, in school and throughout life. http://www.thegeniusworks.com/2017/01/future-education-young-everyone-taught-together

Foucault, M. (1978). *The history of sexuality.* Random House.

Gamble, A. (2012). The United Kingdom: The triumph of fiscal realism?. In W. Grant and G. K. Wilson (Eds.), *The consequences of the global financial crisis: The rhetoric of reform and regulation* (pp. 34–50). Oxford.

Gerstein, J. (2014). Moving from education 1.0 through education 2.0 towards education 3.0. In *Experiences in Self-determined learning* (pp. 83–98). CreateSpace Independent Publishing Platform.

Giannini, S. (2023). *Generative AI and the future of education.* UNESCO.

Gill, K. S. (2018). Artificial intelligence: Looking though the Pygmalion Lens. *AI & Society, 33*(4), 459–465. https://doi.org/10.1007/s00146-018-0866-0

Giroux, H. A. (2013). *On critical pedagogy.* Bloomsbury.

Goodenow, R. K., & Cowen, R. (1986). The American School of education and the third world in the twentieth century: Teachers College and Africa, 1920–1950. *History of Education, 15*(4), 271–289. https://doi.org/10.1080/0046760860150405

Görur, R., Sellar, S., & Steiner-Khamsi, G. (2018). Big data and even bigger consequences. In R. Gorur, S. Sellar, & G. Steiner-Khamsi (Eds.), *The world yearbook of education 2019: Comparative methodology in the era of big data and global networks* (pp. 1–9). Routledge eBooks. https://doi.org/10.4324/9781315147338-1

Grimm, A. (2019). Studying to stay: Understanding graduate visa policy content and context in the United States and Australia. *International Migration, 57*(5), 235–251.

Guan, C., Mou, J., & Jiang, Z. (2020). Artificial intelligence innovation in education: A twenty-year data-driven historical analysis. *International Journal of Innovation Studies, 4*(4), 134–147. https://doi.org/10.1016/j.ijis.2020.09.001

Haenelein, M., & Kaplan, A. (2019). A brief history of artificial intelligence: On the past, present, and future of artificial intelligence. *California Management Review, 61*(4), 5–14. https://doi.org/10.1177/0008125619864

Hans, N. (1958). *Comparative education.* Routledge.

Hayhoe, R., & Mundy, K. (2008). Introduction to comparative and international education: Why study comparative education? In K. Mundy, K. Bickmore, R. Hayhoe, M. Madden, & K. Madjidi (Eds.), *Comparative and international education: Issues for teachers* (pp. 1–22). Teachers College Press.

Hogarth, I. (2018). AI nationalism. https://www.ianhogarth.com/blog/2018/6/13/ainationalism

Holmes, W. (2023). AIED—Coming of age? *International Journal of Artificial Intelligence in Education.* https://doi.org/10.1007/s40593-023-00352-3

Holmes, W., Persson, J., Chounta, I.-A., Wasson, B., & Dimitrova, V. (2022a). Artificial intelligence in education: A critical view through the lens of human rights, democracy and the rule of law. https://rm.coe.int/artificial-intelligence-and-education-a-critical-view-through-the-lens/1680a886bd

Holmes, W., Porayska-Pomsta, K., Holstein, K., Sutherland, E., Baker, T., Shum, S. B., Santos, O. C., Rodrigo, M. T., Cukurova, M., Bittencourt, I. I., & Koedinger, K. R. (2022b). Ethics of AI in education: Towards a community-wide framework. *International Journal of Artificial Intelligence in Education, 32,* 504–526. https://doi.org/10.1007/s40593-021-00239-1

Hudson, C. (2007). Governing the governance of education: The state strikes back? *European Educational Research Journal, 6*(3), 266–282.

Huk, T. (2021). From education 1.0 to education 4.0-challenges for the contemporary school. *The New Educational Review, 66,* 36–46.

International Artificial Intelligence in Education Society. (2024). International AIED society. https://iaied.org/about

Inverardi, P. (2022). The challenges of human dignity in the era of autonomous systems. In H. Werthner, E. Prem, E. A. Lee, & C. Ghezzi (Eds.), *Perspectives on digital humanism* (pp. 25–29). Springer.

Isaac, W. S., Mohamed, S., & Png, M.-T. (2021). Decolonizing AI: Algorithmic oppression is rooted in the colonial project. *Boston Review.* https://www.bostonreview.net/forum_response/decolonizing-ai/

James, K. (2005). International education: The concept, and its relationship to intercultural education. *Journal of Research in International Education, 4*(3), 313–332.

Jobin, A., Ienca, M., & Vayena, E. (2019). The global landscape of AI ethics guidelines. *Nature Machine Intelligence, 1,* 389–399.

jules, t., & Arnold, R. (2021). Constructing global citizenship education at the regional level: Regionalism and Caribbean citizen education. *Globalisation, Societies and Education, 19*(4), 393–404. https://doi.org/10.1080/14767724.2021.1911630

Kaletsky, A. (2011). *Capitalism 4.0: The birth of a new economy in the aftermath of crisis.* PublicAffairs.

Kandel, I. L. (1955). National and international aspects in education. *Harvard Educational Review, 1,* 5–15.

Kaplan, A. (1964). *The conduct of inquiry.* Chandler Publishing Company.

Kazamias, A., & Massialas, B. G. (1982). Comparative education. In H. E. Mitzel (Ed.), *Encyclopedia of educational research* (5th ed., pp. 309–317). Free Press.

Kelly, G. P., & Altbach, P. G. (1986). Comparative education: Challenge and response. *Comparative Education Review*, *30*(1), 89–107. http://www.jstor.org/stable/1188270

Khe Foon, H. E. W., & Chung Kwan, L. O. (2018). Flipped classroom improves student learning in health professions education: A meta-analysis. *BMC Medical Education*, *18*(1), 38. https://doi.org/10.1186/s12909-018-1144-z

Kishimoto, A., Régis, C., Denis, J., & Axente, M. L. (2024). Introduction. In C. Régis, J. Denis, M. L. Axente, & A. Kishimoto (Eds.), *Human-Centered AI a multidisciplinary perspective for policy-makers, auditors, and users edited* (pp. 1–12). CQR Press.

Klees, S. J. (2017). Quantitative methods in comparative education and other disciplines: Are they valid? *Educação & Realidade*, *42*(3), 841–858. https://doi.org/10.1590/2175-623664816

Kurzweil, R. (2005). *The singularity is near: When humans transcend biology*. The Viking Press.

Labaree, D. F. (1997). Public goods, private goods: The American struggle over educational goals. *American Educational Research Journal*, *34*(1), 39–81.

Laney, D. (2001). 3-D data management: Controlling data volume, velocity and variety. Application Delivery Strategies. *META Group Inc.* https://blogs.gartner.com/doug-laney/files/2012/01/ad949-3D-Data-Management-Controlling-Data-Volume-Velocity-and-Variety.pdf

Lather, P. (2013). Methodology-21: What do we do in the afterward? *International Journal of Qualitative Studies in Education*, *26*(6), 634–645. https://doi.org/10.1080/09518398.2013.788753

Lather, P., & St. Pierre, E. A. (2013). Post-qualitative research. *International Journal of Qualitative Studies in Education*, *26*, 629–633. https://doi.org/1 0.1080/09518398.2013.788752

Law, J. (2011). *What's wrong with a one-world world. Presented to the Center for the Humanities*. Wesleyan University. http://www.heterogeneities.net/publications/Law2011WhatsWrongWithAOneWorldWorl d.pdf

Lee, E. (2022). Are we losing control? In H. Werthner, E. Prem, E. A. Lee, & C. Ghezzi (Eds.), *Perspectives on digital humanism* (pp. 3–8). Springer.

Lee, E. A. (2020). *The coevolution: The entwined futures of humans and machines*. MIT Press.

Lewison, M. (2002). Taking on critical literacy: The journey of newcomers and novices. *Language Arts*, *79*(5), 70–77.

Life Magazine. (1970). *Meet Shaky, the first electronic person: The fascinating and fearsome reality of a machine with a mind of its own* (pp. 59–68). Life Magazine.

Liveley, G., & Thomas, S. (2020). Homer's intelligent machines: AI in antiquity. In S. Cave, K. Dihal, & S. Dillon (Eds.), *AI narratives: A history of imaginative thinking about intelligent machines* (pp. 25–48). Oxford University Press. https://doi.org/10.1093/oso/9780198846666.003.0002

Loukides, M., & Lorica, B. (2016). *What is artificial intelligence?* O'Reilly Media, Inc.

Lövbrand, E., Beck, S., Chilvers, J., Forsyth, T., Hedrén, J., Hulme, M., Lidskog, R., & Vasileiadou, E. (2015). Who speaks for the future of Earth? How critical social science can extend the conversation on the Anthropocene. *Global Environmental Change, 32,* 211–218.

Makrides, G. (2019). *The evolution of education from Education 1.0 to Education 4.0: Is it an evolution or a revolution.* Beer Sheva.

Malakhov, A. (2020). Data mining and predictive analytics in digital education: Lessons we can learn from big data that are often discarded. In t. d. jules & F. D. Salajan (Eds.), *The educational intelligent economy: Artificial intelligence, machine learning and the internet of things in education* (pp. 143–159). Emerald Publishing Limited.

Marginson, S., & Mollis, M. (2001). The door opens and the tiger leaps. Theories and reflexivities of comparative education for a global millennium. *Comparative Education Review, 45,* 581–615. http://dx.doi.org/10.1086/447693

Maslej, N., Fattorini, L., Brynjolfsson, E., Etchemendy, J., Ligett, K., Lyons, T., Manyika, J., Ngo, H., Niebles, J.C., Parli, V., Shoham, Y., Wald, R., Clark, J., & Perrault, R. (2023). *The AI Index 2023* Annual Report. Stanford University.

Mbembe, A. (2017). *Critique of Black reason.* Witwatersrand University Press.

McCarthy, J., Minsky, M. L., Rochester, N., & Shannon, C. E. (2006). A proposal for the Dartmouth summer research project on artificial intelligence, August 31, 1955. *AI Magazine, 27*(4), 12–14. https://www.aaai.org/ojs/index.php/aimagazine/article/view/1904/1.802

McDonnell, S., & Regenvanu, R. (2022). Decolonization as practice: Returning land to indigenous control. *AlterNative: An International Journal of Indigenous Peoples, 18*(2), 235–244. https://doi.org/10.1177/11771801221100963

Means, A. (2018). *Learning to save the future: Rethinking education and work in an era of digital capitalism.* Taylor and Francis.

Merriam, S. B., & Tisdell, E. J. (2016). *Qualitative research: A guide to design and implementation* (4th ed.). Jossey-Bass.

Mhlambi, S., & Tribelli, S. (2023). Decolonizing AI ethics: Relational autonomy as a means to counter AI harms. *Topoi, 42,* 867–880.

Miao, F., Holmes, W., Huang, R., & Zhang, H. (2021). *Artificial intelligence and education. Guidance for policy-makers.* UNESCO.

Mignolo, W. D. (2007). Delinking: The rhetoric of modernity, the logic of coloniality and the grammar of de-coloniality. *Cultural Studies, 21*(2–3), 449–514.

Mignolo, W. D. (2009). Epistemic disobedience, independent thought and decolonial freedom. *Theory, Culture & Society, 26*(7–8), 159–181.

Mignolo, W. D., & Walsh, C. E. (2018). *On decoloniality: Concepts, analytics, praxis.* Duke University Press.

Mitchell, T. (1980). *The need for biases in learning generalizations by the need for biases in learning generalizations* (Rutgers CS tech report CBM-TR-117). Rutgers University.

Mohamed, S. (2018). Decolonial artificial intelligence. *The Spectator.* https://blog.shakirm.com/2018/10/decolonising-artificial-intelligence

Mohamed, S., Png, M., & Isaac, W. (2020). Decolonial AI: Decolonial theory as sociotechnical foresight in artificial intelligence. *Philosophy and Technology, 3,* 659–684. https://doi.org/10.1007/s13347-020-00405-8

Moore, B. (2014). *Elephants in space: The past, present and future of life and the universe.* Springer. https://doi.org/10.1007/978-3-319-05672-2

Moore, J., & Newell, A. (1974). How can Merlin understand? In L. W. Gregg (Ed.), *Knowledge and cognition.* Psychology Press.

Moosavi, L (2020). The decolonial bandwagon and the dangers of intellectual decolonisation. *International Review of Sociology 30*(2): 332–354.

Mora, J., & Diaz, D. (2004). *Latino social policy: A participatory research model* (1st ed.). Routledge.

Muldoon, J., & Wu, B. A. (2023). Artificial intelligence in the colonial matrix of power. *Philosophy & Technology, 36*(80). https://doi.org/10.1007/s13347-023-00687-8

Murphy, J. M., & Largacha-Martínez, C. (2022). Decolonization of AI: A crucial bling spot. *Philosophy & Technology, 35*(102), 1–13.

Nemorin, S., Vlachidis, A., Ayerakwa, H. M., & Andriotis, P. (2023). AI hyped? A horizon scan of discourse on artificial intelligence in education (AIED) and development. *Learning, Media and Technology, 48*(1), 38–51. https://doi.org/10.1080/17439884.2022.2095568

Nilsson, N. J. (2009). *The quest for artificial intelligence: A history of ideas and achievements.* Cambridge University Press.

Noah, H., & Eckstein, M. (1969). *The development of comparative education. Toward a science of comparative education.* Macmillan.

Nolan, A. (2023). Accelerating science could be the most valuable use of AI. *OCED. AI Policy Observatory.* https://oecd.ai/en/wonk/accelerating-science

Nordtveit, B. H., & Nordtveit, F. (2020). The educational intelligent economy and big data in comparative and international education research: A decolonial vision. In t. jules & F. D. Salajan (Eds.), *The educational intelligent economy: Artificial intelligence, machine learning and the internet of things in education.* Emerald Publishing Limited.

O'Neill, S. (2023). A (relatively) brief history of AI. https://www.lxahub.com/stories/a-relatively-brief-history-of-ai

O'Reagan, G. (2016). *History of artificial intelligence: Introduction to the history of computing, undergraduate topics in computer science.* Springer. https://doi.org/10.1007/978-3-319-33138-6_19

OECD. (2023). *Education at a glance 2023: OECD indicators*. OECD Publishing. https://doi.org/10.1787/e13bef63-en

Ought. (2023). *Elicit*. https://elicit.com/.

Ouyang, F., & Jiao, P. (2021). Artificial intelligence in education: The three paradigms. *Computers and Education: Artificial Intelligence, 2*, 100020. https://doi.org/10.1016/j.caeai.2021.100020

Peters, S. (2007). "Education for all?" A historical analysis of international inclusive education policy and individuals with disabilities. *Journal of Disability Policy Studies, 18*(2), 98–108.

Peyron, L. (2021). Esserci, o non esserci? Il dilemma dell'intelligenza artificiale, *Moralia Blog*. https://ilregno.it/moralia/blog/esserci-o-nonesserci-il-dilemma-dellintelligenza-artificiale-luca-peyron

Poellhuber, B., Roy, N., & Lepage, A. (2024). Artificial intelligence in higher education: Opportunities, issued and challenges. In C. Régis, J. Denis, M. L. Axente, & A. Kishimoto (Eds.), *Human-centered AI a multidisciplinary perspective for policy-makers, auditors, and users edited* (pp. 151–162). CQR Press.

PRC Ministry of Education. (2018). Artificial intelligence innovation action plan for institutions of higher education. https://oecd.ai/en/dashboards/policy-initiatives/http:%2F%2Faipo.oecd.org%2F2021-data-policyInitiatives-26851

PRC Ministry of Science and Technology. (2021). Ethical norms for new generation artificial intelligence released. https://cset.georgetown.edu/publication/ethical-norms-for-new-generation-artificial-intelligence-released/

Quijano, A. (2000). Coloniality of power and eurocentrism in latin America. *International Sociology, 15*(2), 215–232. https://doi.org/10.1177/0268580900015002005

Quijano, A. (2007). Coloniality and modernity/rationality. *Cultural Studies, 21*(2–3), 168–178.

Quijano, A. (2016) "Bien Vivir—Between development and the de/coloniality of power". Alternautas (Re) searching development: The Abya Yala Chapter 3(1), 10–23. http://www.alternautas.net/blog/2016/1/20/bien-vivir-between-development-andthe-decoloniality-of-power1

Quijano, A. (2017). Good Living': Between development and the de/coloniality of power. In W. Raussert (Ed.), *The Routledge companion to Inter-American Studies* (pp. 363–371). Routledge.

Ravitch, S. M., & Carl, N. M. (2020). *Qualitative research: Bridging the conceptual, theoretical, and methodological*. Sage.

Resnick, L. B. (1987). The 1987 presidential address: Learning in school and out. *Educational Researcher, 16*(9), 13–54. https://doi.org/10.3102/0013189X016009013

Richardson, C., Oster, N., Henriksen, D., & Mishra, P. (2023). Artificial intelligence, responsible innovation, and the future of humanity with Andrew Maynard. *TechTrends: Rethinking Creativity and Technology in Education*. https://doi.org/10.1007/s11528-023-00921-2

Roll, I., & Wylie, R. (2016). Evolution and revolution in artificial intelligence in education. *International Journal of Artificial Intelligence in Education, 26*, 582–599. https://doi.org/10.1007/s40593-016-0110-3

Rushby, N. (2013). Data-sharing. *British Journal of Educational Technology., 44*(5), 675–676.

Russell, S. (2022). Artificial intelligence and the problem of control. In H. Werthner, E. Prem, E. A. Lee, & C. Ghezzi (Eds.), *Perspectives on digital humanism* (pp. 19–24). Springer.

Russell, S., & Norvig, P. (2021). Artificial intelligence: A modern approach, 4th US ed. aima: сайт. https://aima.cs.berkeley.edu/

Salajan, F. D., & jules, T. D. (2020). Introduction: The educational intelligent economy, educational intelligence, and Big Data. In t. d. jules & F. D. Salajan (Eds.), *The educational intelligent economy: Big data, artificial intelligence, machine learning and the internet of things in education* (pp. 1–32). Emerald Publishing.

Salajan, F. D., & jules, t. d. (2021). A comparative examination of regulated and unregulated big data analytics as (re)makers of complex educational assemblages in the European Union and the Caribbean Community. In A. W. Wiseman (Ed.), *Annual review of comparative and international education 2020* (pp. 149–170). Emerald Publishing Limited.

Salmon, G. (2019). May the fourth be with you: Creating education 4.0. *Journal of Learning for Development, 6*(2), 95–115.

Saltman, K. J. (2022). *The alienation of fact: Digital educational privatization, AI, and the false promise of bodies and numbers.* The MIT Press.

Santos, O. C., Rodrigo, M. T., Cukurova, M., Bittencourt, I. I., & Koedinger, K. R. (2022b). Ethics of AI in education: Towards a community-wide framework. *International Journal of Artificial Intelligence in Education, 32*, 504–526. https://doi.org/10.1007/s40593-021-00239-1

Schiff, D. S. (2022). Education for AI, "Not" AI for education: The role of education and ethics in national AI policy strategies. *International Journal of Artificial Intelligence in Education, 32*(3), 527–563.

Schwab, K. (2016). The fourth industrial revolution: What it means, how to respond. https://www.weforum.org/agenda/2016/01/the-fourth-industrial-revolution-what-it-means-and-how-to-respond/

Sicka, B., & Hou, M. (2023). Dismantling the master's house: A decolonial blueprint for internationalization of higher education. *Journal of Comparative & International Higher Education, 15*(5), 27–43.

Silver, D., Huang, A., Maddison, C. J., Guez, A., Sifre, L., van den Driessche, G., et al. (2016). Mastering the game of go with deep neural networks and tree search. *Nature, 529*(7587), 484–489. https://doi.org/10.1038/nature16961

Smalheiser, N. R., Hanh-Powell, G., Hristovski, D., & Sebastian, Y. (2024). *From knowledge discovery to knowledge creation: How can literature-based discovery accelerate progress in science?* Oecd-Ilibrary.org; Organization for Economic Cooperation and Development. https://www.oecd-ilibrary.org/sites/de74d7a9-en/index.html?itemId=/content/component/de74d7a9-en

Smith, L. T. (1999). *Decolonizing methodologies: Research and indigenous peoples* (3rd ed.). Bloomsbury Academic.

Sora, J. C., & Sora, S. A. (2012). Artificial education: Expert systems used to assist and support 21st century education. *GSTF Journal on Computing, 2*(3), 1–4. https://doi.org/10.5176/2010-3043_2.3.177

Spector, L. (2006). Evolution of artificial intelligence. *Artificial Intelligence, 170*(18), 1251–1253. https://doi.org/10.1016/j.artint.2006.10.009

Sujon, Z. (2019). Disruptive play or platform colonialism? The contradictory dynamics of Google expeditions and educational virtual reality. *Digital Culture & Education, 11*(1) https://www.digitalcultureandeducation.com/volume-11-papers/disruptive-play-or-platform-colonialism-the-contradictory-dynamics-of-googleexpeditions-and-educational-virtual-reality

Sutoris, P. (2022). *Educating for the Anthropocene: Schooling and activism in the face of slow violence.* The MIT Press.

Sylvester, R. (2002). Mapping international education: A historical survey 1983–1944. *Journal of Research in International Education, 1*(1), 90–125.

Sylvester, R. (2003). Further mapping of the territory of international education in the 20th century (1944–1969). *Journal of Research in International Education, 2*(2), 185–204.

Sylvester, R. (2005). Framing the map of international education (1969–1998). *Journal of Research in International Education, 4*(2), 123–151.

Takayama, K. (2018). Beyond comforting histories: The colonial/imperial entanglements of the International Institute, Paul Monroe, and Isaac L. Kandel at Teachers College, Columbia University. *Comparative Education Review, 62*(4), 459–481. https://doi.org/10.1086/699924

Takayama, K. (2020). Engaging with the more-than-human and decolonial turns in the land of shinto cosmologies: "Negative" comparative education in practice. *ECNU Review of Education, 3*(1), 46–65. https://doi.org/10.1177/2096531120906298

Takayama, K. (2022). *Doing southern theory: Shinto, self-negation, and comparative education* (pp. 589–606). Springer EBooks. https://doi.org/10.1007/978-3-030-86343-2_33

Takayama, K., Sriprakash, A., & Connell, R. (2017). Toward a postcolonial comparative and international education. *Comparative Education Review, 61*(S1), S1–S24. https://doi.org/10.1086/690455

Tamburrini, G. (2022). Digital humanism and global issues in artificial intelligence ethics. In H. Werthner, E. Prem, E. A. Lee, & C. Ghezzi (Eds.), *Perspectives on digital humanism* (pp. 83–88). Springer.

Tate, N. (2012). Challenges and pitfalls facing international education in a post-international world. *Journal of Research in International Education*, *11*(3), 205–217.

The Economist. (2019). A brief history—and future—of credit scores. https://www.economist.com/international/2019/07/06/a-brief-history-and-future-of-credit-scores

The Economist. (2023). How scientists are using artificial intelligence. https://www.economist.com/science-and-technology/2023/09/13/how-scientists-are-using-artificial-intelligence

Thomas, G. (2013). *Education: A very short introduction*. Oxford University Press.

Tuck, E., & Yang, W. K. (2012). Decolonization is not a metaphor. *Decolonization: Indigeneity, Education & Society*, *1*(1), 1–40.

Turing, A. M. (1950). Computing machinery and intelligence. *Mind, LIX*(236), 433–460. https://doi.org/10.1093/mind/LIX.236.433

Turnbull, D., Chugh, R., & Luck, J. (2021). Learning management systems: A review of the research methodology literature in Australia and China. *International Journal of Research & Method in Education*, *44*(2), 164–178. https://10.1080/1743727X.2020.1737002

Ulnicane, I., Eke, D. O., Knight, W., George Ogoh, G., & Stahl, C. B. (2021). Good governance as a response to discontents? Déjà vu or lessons for AI from other emerging technologies. *Interdisciplinary Science Reviews*, *46*(1–2), 71–93. https://doi.org/10.1080/03080188.2020.1840220

UNESCO. (2019a). Artificial intelligence in education: Challenges and opportunities for sustainable development. https://unesdoc.unesco.org/ark:/48223/pf0000366994

UNESCO. (2019b). *Planning education in the AI era: Lead the leap*. UNESCO. https://unesdoc.unesco.org/ark:/48223/pf0000370967

UNESCO. (2021a). Artificial intelligence in education: Challenges and opportunities for sustainable development. https://www.gcedclearinghouse.org/sites/default/files/resources/190175eng.pdf

UNESCO. (2021b). *AI and education: Guidance for policymakers*. UNESCO.

UNESCO. (2023a). Generative artificial intelligence education: What are the opportunities and challenges? https://www.unesco.org/en/articles/generative-artificial-intelligence-education-what-are-opportunities-and-challenges?hub=81942

UNESCO. (2023b). Global education monitoring report 2023: Technology in education: A tool on whose terms? UNESCO.

Vallor, S. (2024). Defining human-centered AI: An interview with Shannon Vallor. In C. Régis, J. Denis, M. L. Axente, & A. Kishimoto (Eds.), *Human-centered AI a multidisciplinary perspective for policy-makers, auditors, and users edited* (pp. 13–20). CQR Press.

Vinuesa, R., Azizpour, H., Leite, I., Dignum, V., Domisch, S., Felländer, A., Langhans, S. D., Tegmark, M., & Nerini, F. F. (2020). The role of artificial intelligence in achieving the sustainable development goals. *Nature Communications, 11*, 233. https://doi.org/10.1038/s41467-019-14108-y

Wa Thiong'o, N. (1986). *Decolonising decolonizing the mind: The politics of language in African literature.* Heinemann Educational.

Weizenbaum, J. (1966). ELIZA—a computer program for the study of natural language communication between man and machine. *Communications of the ACM, 9*(1), 36–45. https://doi.org/10.1145/365153.365168

Weizenbaum, J. (1976). *Computer power and human wisdom: From judgment to calculation.* W.H. Freeman and Company.

Werthner, H., Prem, E., Lee, E. A., & Ghezzi, C. (Eds.). (2022). *Perspectives on digital humanism.* Springer.

Williams, G. (2016). Higher education: Public good or private commodity? *London Review of Education, 14*(1), 131–142.

Williamson, B. (2023). The social life of AI in education. *International Journal of Artificial Intelligence in Education.* https://doi.org/10.1007/s40593-023-00342-5

Wiseman, A. W. (2022). Recent developments in the relationship between empirical comparative research on education and neo-institutional theory. *Internationales Jahrbuch der Erwachsenenbildung, 45*(1), 15–38.

Wynter, S. (2003). Unsettling the coloniality of being/power/truth/freedom: Towards the human, after man, its overrepresentation—an argument. *CR: The New Centennial Review, 3*(3), 257–337. http://www.jstor.org/stable/41949874

Zawacki-Richter, O., Marín, V. I., Bond, M., & Gouverneur, F. (2019). Systematic review of research on artificial intelligence applications in higher education—where are the educators? *International Journal of Educational Technology in Higher Education, 16*(39), 1–27. https://doi.org/10.1186/s41239-019-0171-0

Zimeta, M. (2023). *Why AI must be decolonized to fulfill its true potential.* https://www.chathamhouse.org/publications/the-world-today/2023-10/why-ai-must-be-decolonized-fulfill-its-truepotential

INDEX[1]

---

[1] Note: Page numbers followed by 'n' refer to notes.

© The Author(s), under exclusive license to Springer Nature
Switzerland AG 2024
S. M. S. Curtis et al., *The Technological-Industrial Complex and
Education*, https://doi.org/10.1007/978-3-031-60469-0

GPSR Compliance

The European Union's (EU) General Product Safety Regulation (GPSR) is a set of rules that requires consumer products to be safe and our obligations to ensure this.

If you have any concerns about our products, you can contact us on ProductSafety@springernature.com

In case Publisher is established outside the EU, the EU authorized representative is:

Springer Nature Customer Service Center GmbH
Europaplatz 3
69115 Heidelberg, Germany

The manufacturer's authorised representative in the EU is Springer
Nature Customer Service Centre GmbH, Europaplatz 3, 69115 Heidelberg,
Germany. If you have any concerns regarding our products, please
contact ProductSafety@springernature.com

Printed and bound by CPI Group (UK) Ltd, Croydon, CR0 4YY
29/04/2026
02099531-0006